Agro-Meteorological Modelling

-Principles, Data and Applications

Agro-Meteorological Modelling

-Principles, Data and Applications

(With examples drawn from Ireland)

Nicholas M. Holden (Editor)

ΔϚϜΣϹ

Joint Working Group on Applied Agricultural Meteorology
Dublin • Ireland • 2001

Published in 2001 in Ireland by:
The Joint Working Group on Applied Agricultural Meteorology (Agmet)
c/o Met Éireann
Glasnevin Hill
Dublin 9
Ireland

Holden, N. M
Agro-Meteorological Modelling – Principles, Data and Applications
1. Climate–Meteorology–Weather–Agricluture–Mathematical Modelling–Ireland
I. Title, II. Holden, III AGMET

ISBN   0 951155164 (hardback)
        0 951155172 (softback)

Printed in Ireland by Colourbooks Ltd.

Previous Agmet publications

Climate, Weather and Irish Agriculture, edited by T. Keane, 1986

Proceeding of Conference on Weather and Agriculture, edited by T. Keane, 1988

Weather, Soils and Pollution form Agriculture, compiled by M. Sherwood, 1992

The Future of Irish Agriculture–Role of Climate, edited by J. F. Collins, 1992
(proceedings of a conference held at University College Dublin)

Irish Farming, Weather and Environment, edited by T. Keane, 1992

The Balance of Water, edited by T. Keane and E. Daly, 1994
(proceedings of a conference held at Trinity College Dublin)

Agroclimatic Atlas of Ireland, edited by J. F. Collins and T. Cummins, 1996

Weather and Agro-environmental Management, edited by D. McGilloway, 2000
(proceedings of a conference held at The Geological Survey of Ireland)

Contents

Preface

1	Modelling Concepts	1
2	Introduction to Geographic Information Systems, Remote Sensing and Geostatistics	23
3	Meteorological Data–Types and Sources	56
4	Terrestrial Data	80
5	A Comparison of Grass Growth Models	136
6	A GIS based Model of Catchment Surface Water Quality	155
7	Crop and Animal Disease Forecasting and Control–Regional Perspectives	185
8	Weather Forecasting by Numerical Weather Prediction	221
9	An Endnote: The Future Development of Models and Modelling in Agro-Meteorology	245

Acknowledgements

The Agmet group wishes to thanks the following for reviewing the contents of this book: J. Thornley, M. Hough, G. Smillie, G. Mills, G. Lemaire, T. Crawley, S. Laidlaw, D. O'Brien, R. Hammond, E. Goodall, P Lynch and R. Bates.

The financial support of the following organisations is also greatly appreciated: Departments of Crop Science, Horticulture & Forestry, Environmental Resource Management and Agricultural & Food Engineering, University College Dublin; National University of Ireland, Maynooth; Teagasc; Met Éireann; Gouldings Chemicals Ltd; R&H Hall & Co; Zeneca Ltd; BASF (Ireland) and the National University of Ireland (Publication Grant Aid).

Preface

This book is aimed at all those with an interest in application of models to environmental issues, but more specifically to the rural environment. The weather plays a significant role in controlling how man can manage the environment and the consequences of his actions. Since the first AGMET publication: *Climate Weather and Irish Agriculture* (T. Keane, Ed.) in 1986 there have been many changes in the technologies available for assisting land managers. The introduction of the personal computer and the associated revolution in information technology has lead to the development of complex agricultural models. Those working in environment-related areas (farmers, planners, scientists, engineers) need to be computer literate, and to have a good understanding of how models can be used. For the newcomer to the subject, teasing out the basics can be a difficult task. Trying to link fundamental theory with applications published in the scientific literature can be very difficult. This book aims to make life easier by providing: (i) an explanation of the fundamental principles of scientific modelling with respect to the types of models that can be constructed, how to ensure they are scientifically based and how to go about building and testing them; (ii) an overview of some of the important tools the aspiring modeller should be familiar with such as GIS, remote sensing and geostatistics – such tools are now so fundamental to environmental management that anyone interested in this subject should be aware of how they operate and what they can do; (iii) information about the types, sources and availability of data – all models require data, and the quality of the output is wholly dependent on the quality of the initial data used – we will look at where data comes from and how it can be manipulated when it is not initially ideal for the task in hand; and (iv) illustrations of some of the ideas presented with practical examples from Irish research. The context of the book is Irish (the Island of Ireland) simply because that is the region where the contributors work, but the material covered is applicable to people working anywhere in the world. As a final note in this introduction, it must be pointed out that this book is not a complete treatise on the subject of agro-meteorological modelling. There are obvious omissions (such a climate change impacts); however the material that is presented should lead a beginner through the modelling process from fundamental principles, through data requirements to application. There is a broad body of experience contained within the covers of this book, and we hope you will find the content as useful when reading as we have found the experience of its compilation.

ΔSMEC Dublin, 2001

Chapter 1 Modelling concepts

N. M. Holden
Department of Agricultural and Food Engineering
University College Dublin, Ireland

1.1 What is a model?

A model can be defined (Concise Oxford Dictionary) as:

> a simplified...description of a system etc., to assist calculations and predictions.

A good way of thinking about a model is as "a simplified representation of reality" (Thomas and Huggett, 1980). A model is a way of dealing with a complex system in such a way that it becomes manageable and more readily understood. While a model can be thought of as a tool for prediction, it is also a means to evaluate our understanding. These two concepts lead to the idea that there are various types of model, and more than one approach to modelling a system. In this book we will discuss the principles of agro-climatic/meteorological modelling and the data required. This chapter will consider the basic principles of modelling and the various types of model that can be constructed. The reader should be aware that the categories and definitions presented in this chapter are not definitive; rather, they are presented to provide working definitions of terms likely to be encountered when reading the wider literature on modelling and its applications.

There are three principal types of model that can be identified. The *scale model* is a physical model of reality. An *iconic* scale model differs only in size from reality, while an *analogue* scale model differs in size and some other aspects of the system being modelled. A plastic model of a jet aeroplane is an analogue scale model because it looks like the real thing, but is made of plastic. Such a model is a simplification of reality because it does not have all the control systems and an engine, but could still be used for prediction and calculations. The scale model, like all the models discussed in this book should only be used for the purpose for which it is designed, and should not be "stretched" too far. A good example of this would be the architect's three dimensional cardboard model of a proposed building. Such a model would be suitable for evaluating visual impact on a streetscape, but would be totally unsuitable for assessing the air-conditioning requirements for the building.

The *conceptual model* is a list (or diagram) of linked ideas important to the issue being modelled, and their possible interactions. The conceptual model is perhaps the most important type of model because it defines the simplifications that permit a system to be represented. In Chapter 5, an analysis of three approaches to the same problem (grass growth) are presented to illustrate how the conceptual model defines the later stages of agro-meteorological modelling. The conceptual model can lead to the building of an analogue scale model, but is more important as the stepping stone to the final type of model, the *mathematical model*.

The mathematical model is a quantitative model constructed by applying the formal logic of mathematics to a conceptual model. France and Thornley (1984) define a mathematical model as: "…an equation or set of equations which represent the behaviour of a system." It is the quantitative nature of mathematical models that makes them so desirable to the scientist. If a mathematical model can be shown to work well at representing reality as we know it, then it can be used to predict future scenarios quantitatively (given that no underlying assumptions are abused). The conceptual model is used to decide what is important and what can be left out, and then forms the basis of the mathematical model where equations are used to link the component parts. In Chapter 5, three different mathematical models describe the same basic concept of predicting grass growth; in Chapter 6 some of the underlying models of a catchment decision support system are defined; in Chapter 7 a number of mathematical models developed for quantitative disease and pest management are discussed.

We can think of the building of a model as the abstraction of those parts of complex reality relevant to a problem, or put another way, the progressive simplification of reality until it becomes manageable (Thomas & Huggett, 1980).This implies that the modelling process is a circular process, not a linear one (Thomas & Huggett, 1980). We start with a problem about which we formulate a hypothesis, define the important variables, underlying assumptions and linkages (the conceptual model), choose suitable mathematical equations to link the variables (the mathematical model), undertake any necessary calibration, make some predictions and then test the predictions against measured data. Throughout the process of modelling a system, the ideas start in the head of the researcher, progress to paper in the form of linked statements, possibly become realised as a scale or analogue model, and usually end up as a computer program. The medium in which the model is expressed evolves with the model. There are examples of this progression throughout this book.

Once we have evaluated the model we can either conclude that it is a good model (i.e. it predicted well) or it is a poor model. If the model is good we can accept it, but if the model is poor, we need to re-evaluate: (1) is the hypothesis sound?; (2) are all assumptions adhered to?; (3) is the mathematical formulation correct?; (4) are the evaluation data applicable for comparison with the model output?; (5) are any relevant variables missing? and (6) is the computer program correct? If at any of the five steps (in order) we conclude there is a problem, then it needs to be removed before continuing. If we are dissatisfied at the end of this process then we reject the model and go back to evaluate the original idea from which the hypothesis was derived.

An alternative situation encountered when modelling a system is that we try to apply a model that someone else formulated. If a model seems applicable to a particular problem, and is widely used in the scientific community then it seems reasonable to try to use it. In this case it is essential that great care is taken to ensure that the model addresses the problem at hand, and that the underlying assumptions of the model are understood and will be met by the data being used. Users should also be aware that some more complex mathematical formulations can give unexpected results when used with values outside the boundary conditions for which the model was designed. If a model is being applied to a new purpose or even a new geographical area then it may be necessary to calibrate and test the model against local data before it is used for scenario testing.

1.2   The relationship between models and research—why are models developed?

Having considered the meaning of model and modelling in a scientific context, it remains to justify modelling as a valid scientific technique. The main reason that models are developed is that they facilitate advancement of knowledge. The circle of model development provides a robust framework for increasing our understanding of how a system works. In this sense, modelling can be thought of as a basic research activity. In a wider context, models also facilitate social and economic gain because they permit us to predict the future and provide intelligent tools for management of systems. In this sense modelling can be thought of as an applied research activity. In the next subsections we consider the application of science and modelling to basic research (including experimental design), through applied research and into the field of management.

1.2.1   Basic research

Basic research is the scientific activity that underlies all advancements in knowledge. A researcher identifies a problem or area of ignorance, estimates whether it is likely that progress can be made with available resources, sets the boundaries to the problem, formulates a hypothesis or monitoring scheme and then collects data. When formulating a hypothesis it is likely that a conceptual model will be developed, and in making sense of the data it is likely that a mathematical model will be implemented to be tested against measurements. In this way the model helps the basic understanding of the researcher, and later simplifies the system to enable communication of information to other interested parties. The mechanistic models defined in Chapter 5, and the building blocks of the decision support system outlined in Chapter 6 are examples of the relationship between modelling and basic research.

There is an intimate relationship between basic agricultural research and experimental modelling because, at this level, experimentation is the primary tool for testing hypotheses. The use of empiricism (observing changes in a response variable with adjustments in a controlling variable) in basic research is also closely linked to modelling because model development is frequently approached from the point of view of experimental design. Additionally, data used for the testing of models are frequently derived from results of experimentation. In order to ensure that model testing is reliable, the available data should be derived from well designed experiments, or specifically designed sampling strategies. A further consideration is that models can be used to enable researchers to improve their experimental design by providing insight into how a system behaves. Spector (1981) suggested that the following important elements should be evaluated when designing an experiment: (1) the *variables* involved; (2)

measurement techniques used; (3) assessment of *measurement error*; (4) *reliability* or consistency of measurement; (5) *validity* of measurements undertaken; (6) *control* of the experiment; (7) *confounding* variables; and (8) degrees to which results can be *generalised*.

The distinction between basic and applied research is fuzzy. For practical purposes the exact boundary is usually unimportant. However in an increasingly market-driven research environment, much research has to be financially as well as intellectually justified. As a consequence more and more research is falling under the applied context, with basic research only being pursued when an undeniable gap in knowledge has to be filled. Such an approach is probably not very efficient.

1.2.2   Applied research and development

Applied research generally develops an existing understanding of specific parts of a system by integration and enhancement such that an important step in controlling the system can be achieved. Models play an important role in applied research because they represent a distillation of knowledge in a generalised form. Chapter 6 illustrates the development of a system of models (i.e. applied research) that can be used to help manage a river catchment system.

Development is the process of making significant technological advances and occurs when all the basic knowledge is available to achieve a goal but needs to be shaped in a suitable manner. For most purposes the distinction between applied research and development is not necessary because of the overlap created by unforeseen gaps in knowledge (hence the common term R & D). With respect to models and modelling, development is concerned with devising methods of management, control and future prediction with a view to generating economic or social gain.

1.2.3   Management and Control

The development of control systems for management purposes is very similar to model building in that a conceptual model of the process has to be formulated and all interactions defined before control can be implemented. In this context, control implies more than just predicting what will happen; it suggests that physical action will be based directly on the information provided by the control system. In the realm of agricultural management we can consider a control system and a management model to be one and the same thing. Control systems can be thought of as decision support systems in which a series of sub-models are integrated to permit holistic management of a farm. In this context, the weather, and more importantly prediction of the future weather is significant in dictating what actions should be taken, and when.

Models are frequently used for management at many scales. An individual farmer may rely on information from agricultural advisers on potential pollution levels, pest and disease spread, yields and market values. Many of these variables are forecast using models that have been shown to be reasonably accurate predictors. Furthermore, day-to-day activity is likely to be dictated by weather forecasts, many of which rely on atmospheric modelling. Thus the management of agricultural systems is highly dependent on the development of reliable weather models.

1.2.4   Future prediction and scenario testing

One of the most widely used applications of models is the prediction of the future by scenario testing. Scenario testing can also be used to explain the past. Economic forecasting and estimates of the effects of elevated atmospheric carbon dioxide are just two applications of modelling that are commonly discussed in popular media. While such applications are important they are also potentially the most dangerous. The development of chaos theory arose from meteorological modelling, from which it can be seen that small differences in the starting conditions can result in widely varying results. It should also be noted that great care is required when interpreting the output of a model if it has been operating with extreme input data. In the case of long-term future scenario testing (e.g. the sustainability of an agronomic system with elevated atmospheric CO_2), we often have to estimate the likely climatic conditions that will occur, and therefore the prediction can only be as good as the estimate of what might be going to happen.

If we consider the model development process outlined earlier there are two important steps which need to be addressed carefully when using a model for predictive purposes: (1) are the data being used for the scenario testing (i.e. predicting what will happen given a range of conditions) compatible with the assumptions of the hypothesis and mathematical formulation of the model being used? and (2) does the model require any specific calibrations involving variables that are being estimated, and if so what is the model's sensitivity to these calibrations? When a value for a variable has to be estimated, it is common practise to run the model with a range of values to start with, in order to establish a good starting point. A seemingly similar, but subtly different requirement of many models is the judicious adjustment of empirical coefficients ("fitting" parameters) in order to operate. In this case, the number used does not necessarily represent any real physical property, rather it allows the model to operate mathematically. An example in soil water modelling is the use of intercept and gradient parameters derived from water characteristic curves (Campbell, 1974), which are easy to measure in the laboratory. These enable estimates of unsaturated hydraulic conductivity to be made as a function of water content or tension, a parameter difficult to measure in the laboratory. If such input variables are required for a model, but have to be estimated for scenario testing, then effort should be expended in testing model sensitivity to the values used in order to establish how important any estimation errors might be. Semenov and Porter (1995) identified a number of difficulties when making crop prediction involving climate data. The main problem they identified was the uncertainty inherent in predicting what the weather will be like in the future. However they also identified problems with the non-linear response of a crop to weather and the fact that in general, only average weather variables are used in predicting the future climate, with little account being taken of the variability that will happen.

1.2.5   Economic and social gain

It might well be suggested that the driving force behind applied models for future prediction is economic gain, or from a social perspective, the management of social deprivation. There is a tendency to use models as a tool to help extract society from a problem, rather than as tools to foresee problems that will then be avoidable. The cynical might suggest that social management will tend to follow somewhere behind economic gain. In agro-meteorology the concern is understanding the role of weather and climate in agricultural (and related) production with a view to producing enough food to feed populations, making money for the farmer, and trying to limit the environmental impact of agronomic systems to acceptable levels. The model is a central concept in this activity. However, agro-economic models developed by economists have tended to ignore meteorological variables, although it is a relatively easy task to integrate an economic performance spreadsheet with a scenario testing model to predict the financial consequences of management at both the farm and regional scales.

1.3   Types of mathematical model

1.3.1   Empirical and mechanistic models

An *empirical model* is one based entirely on observed relationships and not on predetermined theory. When an empirical model is formulated mathematically, the equations used are not based on any inherent understanding of the mechanisms by which the independent variable (cause) influences the dependent variable (effect). It is understood rather that one variable is influencing the other in some unknown manner. The simplest form of empirical model is the linear regression equation where one variable is used to predict another, and being a statistical model does not necessarily imply cause and effect, merely correlation. Empirical models can be far more complex than a single equation; the N-cycle model developed to predict nitrogen dynamics associated with grassland beef production (Scholefield *et al.*, 1991) is empirical because most links are described by regression equations, but is far from being a simple model.

A *mechanistic model* is one that includes an element of understanding or explanation of the system being modelled. A mechanistic model generally requires a two layer approach whereby predictions at the upper level (e.g. plant/crop growth rates) are made using causal relationships and quantities at the lower level (e.g. photosynthesis, nutrient uptake or transpiration, in the case of plant growth). Such descriptions will tend to be incomplete (but ultimately expandable), so the modeller has to place limits on the extent of the model if it is to remain manageable and usable. The lower level in a mechanistic model may comprise empirical relationships, each of which is included because it represents our understanding of how the system works. It is also possible that a theoretical basis for the lower level can be derived from an even lower level, however the limits of the model in this case should be set by manageability. There is no point in developing a model that requires so much data to run that it can never be used. It is essential to evaluate a mechanistic model with respect to the design objectives. Therefore, a clear definition of the requirements and scope of the model should be made at the outset. Evaluation should be undertaken for the prediction (upper level output) and for the assumptions (lower level "mechanisms"). In general mechanistic models are good research tools because they encourage consideration of the system in a manner not possible with a statistical regression. Ultimately, due to the nature of scientific research, all mechanistic developments are based on empiricism at the outset but there will be a tendency for evolution as understanding increases.

In general an empirical model formulated to describe a particular system will tend to work better than a mechanistic model because it will be site specific. However once an empirical model is applied at the extremes of the data range used to formulate it, or to a similar related system, it can rapidly deteriorate in quality of output. As long as assumptions are not stretched too far, a mechanistic model should not degrade as much.

1.3.2   Static and dynamic models

A *static model* is one which does not have time as a variable. Any time dependency is ignored on the assumption that the rate of change in the system is so slow that it is irrelevant (i.e. the system is at or near equilibrium), or is so short relative to environmental change that the environment can be considered constant (France and Thornley, 1984). Most agro-meteorological models are dynamic but may have internal components that are static such as prediction of hydraulic conductivity from water characteristic curves in soil water models, or light interception by a crop canopy in a crop growth model.

A *dynamic model* is one that contains time as an explicit variable and therefore is expressed as a differential equation where some property of the system changes with time. In the case of the Burns model used in the later example in this chapter, there were no differential equations used in formulating the mathematical model but there was still a time component because each model step represented a finite fragment of time. The conceptual model could however, have been formulated mathematically using differential or difference equations that would have included time (e.g. Equation. 1.1)

1.3.3   Deterministic and stochastic models

Deterministic models predict a definitive outcome for a given set of initial conditions. There is no distribution of outcomes. For a set of input values, the model predicts only one outcome. *Stochastic models* include the concept of distributions associated with some or all inputs and outputs; thus the probability of an outcome can be estimated. What this means is that for a given set of input values there is a range of output values within which the true value lies, or which encompasses the range likely to be found in nature. The type of model used is often linked to the uncertainty associated with the system being modelled. A situation with no clearly defined laws (possibly rainfall rate, total volume and distribution pattern) is not well suited to deterministic modelling: a given weather situation will not always yield a given rainfall distribution. The complexity of stochastic models tend to weigh against their being developed unless no other way is possible. One way around this is to look at predicting trends (rather than absolutes) where a simple law may be sufficient to define the likely outcome. (Whether this would constitute an empirical or mechanistic approach is open to debate.) Any system where chance plays a significant part is poorly suited to deterministic modelling. Having made the earlier statements it should also be noted that for some tasks there is no reason why the two techniques cannot be combined; the distinction between these types of model is largely a convenience for explanatory purposes. Stochastic models, while developed in research environments, tend towards an inherent suitability for management purposes.

1.3.4   Normative models

A *normative model* is one which starts from assumptions of what the behaviour of a system ought to be, and compares the output with reality in order to assess those assumptions. Such models are not common in the physical sciences and are more readily found when studying human systems. An example of a normative type model is the "normal soil" as defined by the Soil Survey Staff (1951) which is used to assist in the classification of soils. A "normal soil" is defined as: "…one having a profile in equilibrium or nearly in equilibrium with its environment, developed under good but not excessive drainage from parent material of mixed mineralogical, physical, and chemical composition, and expressing the full effects of the forces of climate and living matter. The typical representatives of the zonal great soil groups are normal soils." The "normal soil" is effectively used as a model of the ideal soil in an area. It is used as the basis for comparison and evaluation of other soils found in the area. The other soils are evaluated with respect to how they differ from the normal and thus the influence of environmental factors can be estimated and evaluated. The "normal soil" can form the basis of a hypothesis regarding soil development at a location and is therefore being used in exactly the same manner as a mathematical model. It should be noted that a "normal" soil might not actually exist.

1.4   Scales in modelling

Scale generally has two meanings when used in conjunction with models: (1) referring to the type of measurement rule that is used to acquire and report data (directly related to the units of measurement) and; (2) the spatial magnitude represented by a model (i.e. point, field, farm, catchment, region, country, continent, world).

1.4.1   Measurement scales

There are four principal measurement scales. A *nominal* scale works by simply numbering objects 1, 2, 3, . . . , n, where the value assigned is of no mathematical relevance, i.e. the jersey numbers for a soccer squad at most indicate a possible position on the field. It can well be argued that nominal scales are not measurement systems but classifiers. All members linked by a nominal scale must have at least one attribute in common by which they can be compared. Classification in itself is no better or worse than a scale of measurement. The underlying reason for modelling is to develop an understanding of the world based on theory and laws. To achieve this it is usually necessary to establish classes of objects about which generalisations may be made (Johnston, 1976). Nominal scaling can be developed further by counting object numbers in a class and thus developing frequency information such as number of wet days in a period of time. An *ordinal* scale is one where ranking can be applied by magnitude, without numerical value, with an irregular interval. For example, the order of streams and rivers based on the number of confluences is a description by ordinal scale because magnitude is implied but not absolute. *i.e.* a 2nd order stream results from the joining of two 1st order streams, a 3rd order from the joining of two 2nd order etc.., but two second order streams might be completely different in terms of their flow characteristics. Such information is not conveyed in the scale used. When the magnitude of the step between objects is known then an *interval* scale can be used. This means that the ratio between two points on the scale can be calculated; however with an interval scale the ratio is dependent on the unit of measurement because the zero point is arbitrarily set. When there is an absolute zero value for a scale it becomes a *ratio* scale and the ratio between two points will be constant regardless of the units of measurement. The classic example of the interval scale is temperature measurement where the ratio between two temperatures expressed in Celsius and Fahrenheit will be different due to each unit system having a different zero point. The Kelvin scale has an absolute zero point and hence is a ratio scale, as are scales of mass and velocity.

1.4.2   Spatial scale

The spatial scale of a model is determined largely by the underlying assumptions. There are two points of view: firstly that of dimensionality (i.e. one, two or three dimensional) and secondly the physical extent of the dimensions represented (i.e. measured in mm, cm, m, km). In the case of a soil–water movement model scale can be 1-, 2- or 3-dimensional. A 1-dimensional model assumes that the predominant direction of water movement is along a single dimension and that the others can be ignored in order to simplify the system. Such a simplification might be acceptable for modelling water movement through a laboratory soil core or a level field with little spatial variation, but to model water movement in a sloped soil, a second dimension would be required due to the untenability of a 1-dimensional model. If there was significant variation in soil type and properties on the slope then a three dimensional soil concept would be required. Similar dimensional simplification can be found when models of whole catchments are studied. Chapter 6 illustrates how a number of simple models are combined to represent a complex, 3-dimensional system. The assumption of reduced dimensions is often used to simplify calculation to proportions manageable by desktop computers thus permitting the implementation of the model in a practical manner.

Spatial scale also concerns the magnitude of the system being modelled. To use the example of soil water modelling again, one of the most common mechanistic techniques utilises equations to describe water flow in a Darcian flow field (i.e. a porous medium in which Darcy's Law applies: the movement of water is proportional to the driving force). In this case, there is an underlying assumption that the volume being modelled is equal to or larger than a "representative elementary volume" (Bear, 1972) which is the minimum volume of the porous medium that is required to average out small-scale variability. It is not possible to readily quantify the representative elementary volume, but it can be equated with the minimum volume for two columns of soil that have similar properties and exhibit similar behaviour with respect to water movement even though they both have slightly different internal structures. The columns would be considered the same at the *macro* scale, but different at the *micro* scale. Any model based on these assumptions has a minimum spatial extent, and in fact a maximum as well; however the definition of the upper boundary may not be so clear, but would be related to the magnitude of spatial variability within the soil of concern. To take an example from meteorology, a general circulation model is of little value when studying variation in grass production in a specific valley because the spatial assumptions of the general model cannot accommodate the specifics of the task at hand. For example, compare the data requirement for the grass models presented in Chapter 5 with the output of a numerical weather prediction model as discussed in Chapter 8. Care should always be taken to ensure that the underlying spatial assumptions of the model being applied are compatible with the issue being addressed.

1.5   How are models developed?

In defining "model" we briefly considered the stages of model building. Now, each stage will be considered in some detail, using as an example of modelling the movement of soluble, unadsorbed anions (e.g. chloride and nitrate) in a fallow, freely drained soil (Burns, 1974). The model chosen is relatively simple, but the original was formulated in a manner that is not so common today. For example the model is dynamic, but time is only implied in the equations, and the use of mixed, non-SI units is unappealing. The model is described using the original terms of the author, in a more modern notation, and an example is given of an alternative conceptual and mathematical formulation using differential equations.

1.5.1   Problem definition and the hypothesis

Problem definition is the most important stage in model development: if it is poor, the model will be of little worth. Spector (1981) considers that at this stage of research it is acceptable to have a "researchable question" which could be something as simple as "what is the relationship among variables?". Such an approach tends however to be poorly regarded. Measuring all possible variables and looking for relationships lacks a certain scientific finesse, partially because a crucial variable may not have been measured or identified, and because this approach ignores any existing knowledge of underlying processes. Starting from a more theoretical stand-point is regarded as the best course of action in most cases. A precisely asked question is more readily answered than a vague one. In the case of the Burns model, the problem was: *to predict how nitrate would redistribute in a field under given seasonal conditions with a view to optimising fertiliser applications both from an economic and environmental point of view.* Burns concluded that previous modelling attempts required too much user manipulation to achieve a good result, and that they did not accommodate dry years when evaporation would be the dominant transport motivator. Burns defined his problem as: "[to develop] a simple simulation model which can be used directly for predicting the redistribution of soluble salts both upwards and downwards in a fallow soil profile."

All inputs required are standard soil science variables and easily obtained. The principle of the model is that, for freely drained soils, after application of excess water, excess water will drain away until field capacity is reached, after which no drainage will occur. Under evaporative conditions, the soil will dry to an evaporative limit from the surface downwards. This means there are maximum and minimum equilibrium water contents for a given depth of soil, controlled by drainage and evaporation respectively.

Beyond the simple definition of a problem lies the formation of a hypothesis. A hypothesis is defined as "supposition made as basis for reasoning, without assumption of its truth, or as starting-point for further investigation from known facts" (*The Concise Oxford Dictionary*). Thus a hypothesis in terms of model building is not necessarily the same as a statistical hypothesis (which at its simplest starts from the stand-point of two populations being the same—the null hypothesis (Norcliffe, 1977)). The purpose of the hypothesis however is the same for both model development and statistical analysis, namely, to ensure that the probability of an incorrect conclusion is minimised. Burns defined the problem of estimating nitrate leaching from a soil and in effect hypothesised that his assumption of a simple mixing-bucket model (one where all chemicals and water within a defined layer or unit can mix freely to equilibrium) would adequately predict what happens in the real world.

1.5.2   Definitions, assumptions and the conceptual model

Having defined a problem, created a hypothesis to address the problem and set the limits of concern, it is then necessary to define all the required variables that will make the model work, and to state any assumptions being made. The definitions of variables and assumptions associated with the formulation of a model are necessary for a third party to evaluate and utilise a model. Without such knowledge it is not possible to evaluate whether a model was developed with bias, and whether the application fits within the constraints of the assumptions. Bias occurs when there is a systematic predisposition towards a result, or a distortion of the results due to factors having been neglected. In a statistical sense, bias occurs when the mean value of samples drawn from a population systematically deviates from the true population mean. In the development of a model, if at the early stages the model builder thinks that some element of the system is not important, then the model may be biased as a result, leading to a systematic misrepresentation in the output. Having considered the necessary variables, stated and evaluated the tenability of assumptions, the next stage of model building is defining the conceptual model.

The Burns model of salt redistribution in soil defines a model structure with a number of constants and variables:
- the total soil profile is considered to be composed of a number of discrete, homogeneous segments (or layers) each characterised by:
  → field capacity (F, cm^3 cm^{-2}) - the maximum water content after free drainage under field conditions with no evaporation (constant)
  → evaporation limit (L, cm^3 cm^{-2}) - minimum attainable water content due to evaporation (sometimes also known as the residual water content) (constant)

→   starting water content (M, cm^3 cm^{-2}) - initial condition at start of model run (variable)

→   anion concentration (A, kg ha^{-1}) - initial condition at start of model run (variable).

• the balance of water added to, or lost from, the soil is the net water flux (X) (variable) which is calculated using:

→   rainfall and/or irrigation (R, cm) (variable, time related input data) with an anion concentration (T, kg ha^{-1}) (variable, time related input data)

→   daily soil water evaporation (E, cm) (variable)

→   where if X > 0, a leaching routine will be followed; if X < 0, an evaporation routine will be followed; and if X = 0, no action will be taken.

• when there is a positive net balance of water added to the soil (i.e. input exceeds evaporative loss) there will be percolation (W$_p$, cm, variable) if the water content of the surface layer exceeds the field capacity. Water moves to the next segment down. This calculation is repeated for each segment until there is either excess water draining from the base of the profile or a layer is reached where the sum of the percolating water and the currently stored water does not exceed the field capacity.

• when evaporation is dominant there will be a negative net balance of water (W$_c$, cm, variable). Water moves to the next segment above. Water is evaporated from the top segment. The maximum water loss from an individual segment is constrained by the evaporation limit. If the evaporative demand exceeds the water available in the top segment then water will be removed until the evaporation limit is reached, and subsequently for lower segments, via those above until the loss is met.

• the anion content of the rainfall and/or irrigation water is used to adjust the anion content of the surface layer in proportion to the volume of water added.

• the time step of the model is 1 day for the duration of the simulation.

As well as defining the model variables it is usually necessary to make some assumptions. These may underlie the proposed mathematical formulation, be related to issues of scale, or be necessary simplifications. It is worth noting that the example model used includes elements that are not common practice in more recent models. The model uses mixed units (e.g. cm^3 and ha) to represent volume and area which are non-SI units, and also "odd" units (cm^{3}cm^{-2}, which appears to equate to cm, but in fact is cm$_{water}^3$ cm$_{land}^{-2}$). Additionally, the model is not conceptualised in terms of "levels" and "rates" as is more common for systems with flows between compartments (Jones and Luyten, 1998).

Burns assumed the following:

• there is no runoff. Infiltration rate is always greater than the rainfall rate. This is a reasonable assumption most of the time because the problem defines the model only for well drained soils. If however, the data being used to run the model included a severe rain storm event, then this assumption may be over-stretched. This would be an example of using a model outside the situation for which it was designed. When using the model it is important that the physical characteristics defining the soil layers are compatible with this assumption.

- it is assumed that no horizontal movement occurs, therefore the model is a one-dimensional approximation of reality.
- when excess water enters a layer there is a complete mixing of anions to equilibrium. This is an assumption sometimes referred to as a "mixing bucket model". The appropriateness of this assumption is related to:
  → the time-step used: if it is too short then an equilibrium could not possibly be used in approximating reality
  → the depth of the segments defined for describing the soil: if the segments are too thick then the assumption would not be tenable
  → the amount of water in the system: if the system is generally short of water, then there would be physical constraints on the mixing of anions
  → the texture and structure of the soil: this is closely related to the assumption of there being no runoff, thus if the soil is of too fine a texture or poorly structured then the two assumptions would be untenable.
- water and anions are assumed to be at equilibrium before any losses from a segment occur. The equilibrium assumptions are necessary to permit a simple mathematical formulation of the conceptual model.
- there are no other mechanisms influencing the available anion concentration in the soil solution such as binding or other chemical reactions. This type of model would not be suitable for phosphorus which is prone to adsorbtion.

The assumptions listed above illustrate the importance of the early stages in model development. Dividing the soil into ideal layers of uniform thickness will not always conform with real soil horizons but is quite acceptable. Also, an ecosystem where water input or losses occur as a single event every 24 hours does not exist. These simplifications permit the formulation of a model that requires little information from the user, is simple to implement (as a computer program) and might provide good results. If we find that the results from the model are not similar to real measurements then we can examine these assumptions. Having stated the assumptions clearly we can easily decide whether the model is suitable for application to a given task. This may not be so easy for a more complex model. We should also take note of any assumptions that underlie the mathematical formulation that is used to implement the conceptual model. If the results of the modelling exercise are good, then we can reasonably say that the assumptions we made are tenable. This is a good reason in itself to use models, because they test our understanding of a system.

The Burns model can be described with a flow diagram (Figure 1.1) which indicates the fundamental properties which have to be described, and the decisions which have to be made. The concept of the soil being described as a series of layers is widely used in soil water modelling, as is the simplification of a one-dimensional system. An alternative description of the system modelled by Burns is using the compartment notation of Forrester (1961) which defines a source (precipitation from the atmosphere, or irrigation water), two sinks (ground water below the soil profile and the atmosphere when evaporation occurs), any number of layers and rates of movement between the layers.

Simplified as only a leaching model the compartments are linked by downward vertical flow rates. The system can then be described by a series of differential equations:

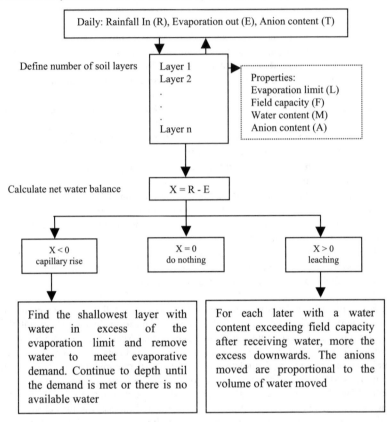

Figure 1.1: Flow diagram of Burns' (1974) conceptual model

$$\frac{dw_1}{dt} = i(t) - q_{1,2}(t)$$

$$\frac{dw_2}{dt} = q_{1,2}(t) - q_{2,3}(t)$$

$$\ldots\ldots$$

$$\frac{dw_n}{dt} = q_{n-1,n}(t) - o(t)$$

1.1

which represent the rate of input from the source, the rate of movement between layers and the rate of output to the sink. While this is an elegant approach to modelling, it requires more information about the soil than the Burns model. Obtaining data for

Equation 1.1 is not as easy as simply defining water contents and limits. These different approaches illustrate the concept of "simplicity of approach". When developing a model, keep things as simple as possible!

1.5.3 Mathematical formulation and programming

The mathematical formulation of a model can be a very complex process even for a simple model. There are a number of issues to consider at this stage. Using the Burns model we can see that, based on the assumptions, it is necessary to perform the calculations that move the water and the anions, and then recalculate the anion concentrations following the mixing-bucket approach. In recent years, software has been developed that permits the mathematical formulation of a model on a computer without the need to use a specific programming language. Examples of such software are *ModelMaker* (Cherwell Scientific Publishing), *CSMP* (Continuous Simulation Modelling Program, California Scientific Software), *Simula* (Norwegian Computing Centre) and *Stella II* (High Performance Systems Inc.). These programs allow the user to define variables, constants and parameters and to use empirical or mechanistic equations to relate one variable with another. Great care needs to be taken when using such programs to ensure that the conceptual model and assumptions are correct, and that the correct techniques for implementing calculations are applied because it is very tempting to use a default method simply because it is available and seems to work.

In the example of Burns' model, all the equations used are simple, but it is necessary to ensure they are performed in the correct order for the model to work. The meteorological variables dictate which operational equation is required to either calculate whether water moves up or down, the volumes of water involved and the consequent movement of anions. We will look at each in the order required to execute the model correctly. The original notation used by Burns is presented. Under "mathematical notation", lowercase letters correspond to Burns' uppercase terms, while the subscript i refers to the segment, and the subscript j refers to the day number. To calculate whether there is a net application or loss of water from the soil surface:

Burns' notation (Burns, 1976)	Mathematical notation	Observations	
$X = R - E$	$x_{1j} = r_j - e_j$	Dictates whether a leaching or evaporation routine will be followed. Only applies to layer 1. [cm]	1.2

where x_{1j} = the net water in/out at the surface layer on a given day, r_j is the rainfall/irrigation and e_j is the surface evaporative demand all on the jth day. Water movement from a segment with excess availability (i.e. where the total available water exceeds the field capacity) is equal to the new water content minus field capacity:

$W_p = X + M - F$	$w_{ij} = x_{ij} + m_{ij} - f_i$	Leaching routine (X>0). [cm]	1.3

where w_{ij} is the water lost (for all layers other than w_1, the surface layer), x_{ij} is the net water into the layer which is equal to $w_{i-1\,j}$, the water leaving the layer above except for layer 1 where it is defined by equation 1.2, m_{ij} is the water content at the start of the time step for layer i on the jth day and f_i is the field capacity of the ith layer. To calculate the fraction of anions transferred downwards by leaching, the ratio of water leaving the segment to the total water available is used (Z_p, Burns' notation):

$Z_p = \dfrac{W_p}{(X+M)}$	$z_{ij} = \dfrac{w_{ij}}{x_{ij} + m_{ij}}$	Anion transfer proportion. [unitless]	1.4

where z_{ij} is the ratio of water leaving segment i on the jth day. Using this ratio the quantity of anions leached (symbol not defined by Burns) is calculated by:

$(A+T)Z_p$	$n_{ij} = z_{ij}.(a_{ij} + t_{ij})$	Quantity of anions leached [kg ha^{-1}]	1.5

where n_{ij} is the quantity of anions leached, a_{ij} is the quantity of anions at the start of the time step, t_{ij} is the quantity of anions in the rainfall/irrigation for layer i on the jth day. If there is no excess water available (i.e. there is no rainfall/irrigation, or the available water is not sufficient to raise the water content to exceed field capacity) then, if the water content of at least one segment is greater than the evaporative limit, the volume available for evaporation (S) is calculated by:

$S = M - L$	$s_{ij} = m_{ij} - l_i$	Capillary rise (X<0) when evaporation exceeds rainfall. [units = cm$_{water}^3$ cm$_{land}^{-2}$]	1.6

where s_{ij} = the water available for evaporation in the layer i on the jth day, and l_1 is the evaporation limit of segment i. Water movement starts from the topmost segment that satisfies the requirement S > 0, and if |X| represents the magnitude of X, then the consequence of the evaporation requirement depends on whether it can be met by water available in that segment:

– if there is enough available water in the segment:

$|X| \leq S, W_c = |X|$	$x_j \leq s_{ij}, w_{ij} = x_j$	[cm]	1.7

– if there is insufficient available water in the segment to match the evaporative demand:

$|X| > S, W_c = S$	$x_j > s_{ij}, w_{ij} = s_{ij}$	[cm]	1.8

If the evaporative demand exceeds the available water in the topmost segment satisfying S > 0 then the remainder of the evaporation requirement is held over to be removed from the next layer down, via the layer(s) above. Once again, to calculate the fraction of anions transferred upwards, the ratio of water moved to the starting water content for the time step is used:

$Z_c = \dfrac{W_c}{M}$	$z_{ij} = \dfrac{w_{ij}}{m_{ij}}$	Same principle as Eqn. 1.3 [unitless]	1.9

Burns then redefined T as the actual quantity of anions lost such that:

$T = A.Z_c$	$n_{ij} = z_{ij}.(a_{ij} + t_{ij})$	Same principle as Eqn. 4. [kg ha^{-1}]	1.10

The final water content of the segment (M_t) after one time step (24 hours) is the starting water content minus that water lost:

$M_t = M - W_c$	$m_{ij+1} = m_{ij} - w_{ij}$	[cm$_{water}^3$ cm$_{land}^2$]	1.11

thus the water content at the start of the next time step (day $j+1$) for each segment is simply the start water content minus the water that is required to meet evaporative demand. In the case of downward percolation the final water content is F (or f_l).

It can be seen that all the equations required for the example model are additions, subtractions, divisions or multiplications. The order is relatively straightforward: for a given time interval, calculate whether there is net gain or loss of water, calculate how much, move the water, calculate the influence on anion concentrations, perform the adjustments, calculate for each subsequent deeper layer until all layers have been processed or there is no requirement to move more water. Measurement error associated with establishing starting conditions and constants can be seen to be important at this stage. The necessary inclusion of division and multiplication can potentially amplify any errors. To illustrate this point, if we incorrectly measured values for two variables which should be 2 and 3, as 2.1 and 3.1, then if they are used in an addition the total resulting error will be 4 % (0.2 ÷ 5). If multiplied the error would be 8.5 % (0.51 ÷ 6).

Having described the mathematical implementation of Burns' model we can consider the type of number that the symbols represent. In a model numbers can represent: (i) constants, like π, (ii) parameters, such as a and b in an equation of the form $y = a + bx$, or fixed values that cannot change in the model, but are variable depending on the situation in which the model is being used: F and L are examples in the Burns' model; (iii) state variables, which define the characteristics of a model element at a particular time: M and A are examples in the Burns model; (iv) rate variables, which indicate the magnitude of change in one variable relative to another: Z_p and E are examples in Burns' model, and; (v) input variables, which are data that "drive" the model through time where each new datum referred to represents a step through time: X, R and T are examples in the Burns model.

1.5.4   Calibration and initialisation of the model

One of the most difficult stages in modelling is obtaining data for the input variables of a model and the subsequent adjustment of parameter values in the calibration process. While this seems superficially easy it often proves not to be. In the example model it is a prerequisite to characterise each layer in the soil with respect to its field capacity, evaporation limit, initial water content and initial anion concentration. In his original presentation of the model, Burns (1974) chose to use 12 segments to describe his soil profile. This meant that he needed 48 data values plus rainfall and evaporation potential data before he could even use the model. The site he used was naturally enough convenient for him with readily available meteorological data. To determine initial water content and anion concentration Burns used 45 cm cores taken at the field site and sectioned into 3.75 cm lengths. To obtain field capacity values he applied excess irrigation, covered with polythene, and then determined water contents by taking core samples after 48 h. The evaporation limit was found by sampling at the end of a 5–6 week dry spell. His weather data came from within 300 m of the test site. The rainfall was measured directly for 24 h periods, but evaporation potential had to be estimated using local information based on values for open water surfaces. This means that even for a relatively simple model, the builder and tester had to rely on an extended dry spell occurring (for evaporation limits), estimating vital input values (evaporation potential), arbitrarily defining the time period for reaching field capacity, and the use of wet chemistry (anion concentration). Imagine the difficulty that arises when trying to apply the model to a location with a rain gauge 5 or 6 km away, no good data on evaporation potential, no long dry periods in the summer and the significant cost of chemical analysis. One conclusion that can be drawn is that outside of a funded research environment, the use of a model has to have a clearly advantageous cost–benefit ratio before it can be countenanced. If a potential user of the model wanted to create data suitable to try a few test runs then field capacity could be estimated from the water characteristic curve (which in turn could be estimated from texture and structure classes), the anion concentration could be estimated from soil survey data for similar sites, and the meteorological data could be from either the nearest measurement sites or interpolated. Techniques for manipulating available data into a form usable by the modeller will be discussed in Chapters 3 and 4. The example applications presented will discuss how problems of model initialisation were overcome.

Calibration is the procedure whereby parameters (and sometimes variables) have their values adjusted by repeated running of the model until the result is close to some predetermined value or curve. Burns was applying his model to a site at a research centre that could be extensively studied, and about which much was known. The advantage of using the model for well known conditions meant that calibration was limited, but if the model were to be applied at an unknown site, then the parameter values and the state variables would have to be adjusted.

1.5.5   Predictions and testing for goodness of fit

Results from Burns' trial of his model indicated that it worked well. Few predicted data points fell outside 2–3 standard deviations of the mean field concentrations. A good linear correlation was possible between predicted and measured data ($r \approx 0.9$). The best-fit line between the predicted and measured data did not differ significantly from 1, which indicated that there was no systematic bias in the model. The prediction of chloride concentrations was better (r = 0.949) than for nitrate (r = 0.918). Burns accounted for this by suggesting that denitrification would have had a role in determining nitrate levels. From this we might conclude that denitrification was not important at this site, since the reduction in the r-value for nitrate was small. The water deficits predicted were the poorest element of the model (r = 0.827). Burns suggests this was partly due to the measurement technique used to acquire the field data. Having evaluated the model and concluded that it is probably a reasonably good model (according to the statistical analysis) we can choose to accept the underlying hypothesis and assumptions. The next stage is to consider how to apply the model for prediction and management purposes. The Burns' model has been used by other modellers because its simplicity and data requirements make it very attractive. Haumann and du Preez (1989) tested the model under irrigation and found that it tended to over-predict nitrate concentrations in a fallow sandy loam soil. They concluded that the model could be used in such soils with reasonable confidence and that it should be used as a convenient alternative to making field measurements. De Neve *et al.* (1998) used the model in conjunction with a temperature dependent mineralisation model (which addressed possible concerns with the original assumption that nitrate was a relatively non-reactive solute) to predict nitrate redistribution in a field from crop residues. They found that the model overestimated leaching in the upper layers and underestimated it in the lower layers of the soil (as did Burns, 1974). They did however conclude that the general pattern of nitrate movement was satisfactory, the modelling efficiency was high, but there was still room for improvement.

In a review of solute leaching models for soils Addiscott and Wagenet (1985) observed that models were rarely tested by someone other than the developer. They offer three possible explanations: (1) models are developed which reflect the specialist interests and environment of the researcher and there may be few potential users with coincidental interests; (2) transfer of models from one computer platform to another may be problematic; (3) a researcher can gain more credit by developing a new model than by testing an old one. Another question that arises when a potential user is evaluating models to apply to a problem is what quantitative criteria should be used that can be universally recognised. Willmott *et al.* (1985), Loague & Green (1991) and Jabro *et al.* (1998) present statistical methods of model evaluation concerned with comparing observed and predicted data. Suggested measures are (for details see Jabro *et al.*, 1998): (i) statistical goodness of fit (i.e. the quality of the linear regression between measured and predicted data)—as used by Burns (1974); (ii) maximum error—the maximum error between any simulated point and its respective measured datum; (iii) root mean squared error—the percentage for the total difference between the simulated and the measured values compared to the mean observed values; (iv) modelling efficiency—a measure between 0 and 1 where 1 indicates a perfect simulation of the measured data, and; (v) coefficient of residual mass—a value of zero indicates that there is no bias in the distribution of simulated values when compared with the measured values. When trying to test a soil water model, Holden *et al.* (1996) encountered the problem that it was not possible to determine the values for a number of state variables because that would have meant destroying the experimental site. To address this issue they inverted the problem.

They obtained sufficient data for soil physical properties from around the experimental site to enable them to define ranges of values (e.g. for bulk density, particle size distribution). In order to test the model they used a statistical sampling method (response surface methodology) to select 30 combinations of values drawn from each variable distribution. The model was then run 30 times with the sampled values and the output compared to field data. It was found that there was no similarity between any of the predicted results and those measured for the test site. If there had been some similarity between the model output and field observations, then the variable values associated with the best prediction could have been assumed to be acceptable for further work. As this was not the case it was assumed that the model could not be reliably applied at the site in question.

1.6   The wheres and whyfors of modelling

Modelling, if applied correctly, has few if any down sides. Most of the disadvantages of models and modelling are related to misapplication, or users and developers who ignore the assumptions associated with either the conceptual formulation or the mathematical implementation of the model. The advantages of modelling and model development are directly paralleled by the principles of good scientific method. Poor scientific development will lead to poor models, but correct development will yield valuable results. Modelling has become particularly powerful with advances in computer technology making calculations a trivial matter once programmed. The objectives of a modelling project are of paramount importance, and time and effort spent in their clear formulation will be repaid with interest. The advantages of modelling are:

- careful consideration of the system, its boundaries, scales and important features are needed
- a hypothesis about the system with assumptions requiring testing must be formulated
- the boundaries of operation have to be set
- future prediction is possible
- a cost effective evaluation of the system can be undertaken
- evaluation of potentially hazardous experiments can be performed with no risk
- an evaluation by sensitivity analysis of the most important factors in a system can aid understanding.

There are downsides to modelling which can easily be avoided if the developer and user are aware of their existence, and if care is taken to always remain aware of why the whole process is being undertaken. The advent of the personal computer, while strengthening the role of modelling, also makes it more prone to problems because it is now much easier to implement models. Things to be wary of include:

- it is easy and sometimes convenient to mistake the model for reality
- models are rarely used beyond their testing stage so their true worth, or lack thereof, is never clearly established

- assumptions are easily ignored as they are often not expressly stated leading to misapplication of techniques or "forcing" a model for convenience
- using without a clear process of hypothesis testing (only applicable to basic and applied research models)
- open ended "playing" with a system (which can seem to be advantageous) can lead to transgressing theoretical boundaries and assumptions, thus creating false problems or situations
- availability of good applicable data for evaluation is often lacking, thus the value of the model is often never really known
- accountability for management decisions can be somewhat removed from the person by blaming the model or its developer for poor outcomes or results.

In the following chapters various aspects of agro-meteorological modelling will be examined. Firstly tools for obtaining and managing geo-spatial data (remote sensing, geographical information systems and geostatistics) will be reviewed (Chapter 2), then the types of available data and their manipulation will be reviewed in Chapters 3 and 4. Some examples of model development and assessment (Chapter 5 (grass) and Chapter 6 (catchment water quality)) are presented, followed by a review of agricultural pest and disease modelling in Ireland (Chapter 7). Finally there is an overview of numerical weather prediction (Chapter 8). The subject areas covered are not exhaustive, nor necessarily definitive; they do however provide illustrations of the steps involved in agro-meteorological modelling and how it can be used for modern land management. The content of the chapters of this book should provide (i) a useful overview of the principles and tools of modelling (ii) an assessment of the type and quality of data available to the modeller, (iii) examples of how models have been, and are being used and (iv) practical illustrations linking existing theoretical textbook material with research literature.

References

Addiscott, T. M. & R. J. Wagenet. (1985): concepts of solute leaching in soils: a review of modelling approaches. *Journal of Soil Science* **36**: 411–424.

Bear, J. (1972): *Dynamics of Fluids in Porous Media*. Dover, New York.

Burns, I. G. (1974): A model for predicting the redistribution of salts applied to fallow soils after excess rainfall or evaporation. *Journal of Soil Science* **25**: 165–178.

Campbell, G. S. (1974): A simple method for determining unsaturated conductivity from moisture retention data. *Soil Science* **117**: 311–314

Doyle, C. J. & J. H. M. Thornley. (1982): An economic model of research and development expenditure. *Oxford Agrarian Studies* **11**: 173–187.

Forrester, J. W. (1961): *Industrial Dynamics*. Wiley, New York.

France, J. & J. H. M. Thornley. (1984): *Mathematical Models in Agriculture*. Butterworths, London.

Haumann, E. J. & C. C. du Preez. (1989): Prediction of nitrate movement in a fine sandy loam soil during the fallow periods with the Burns model. *South African Journal of Plant and Soil Science* **6**: 203–209.

Holden, N. M., A. Rook & D. Scholefield. (1996): Testing the performance of a one-dimensional solute transport model (LEACHC) using response surface methodology. *Geoderma* **69**: 157–173.

Jabro, J. D., J. D. Toth & R. H. Fox. (1998): Evaluation and comparison of five simulation models for estimating water drainage fluxes under corn. *Journal of Environmental Quality* **27**: 1376–1381.

Johnston, R. J. (1976): *Classification in Geography*. Concepts and Techniques in Modern Geography, no. 6. The Institute of British Geographers, London.

Jones, J. W. & J. C. Luyten. (1998): Simulation of biological processes. In R. M. Peart & R. B. Curry (Eds.) *Agricultural Systems, Modeling and Simulation*. Marcel Dekker Inc. New York. p. 19–62.

Loague, K. & R. E. Green. (1991): Statistical and graphical methods for evaluating solute transport models: overview and application. *Journal of Contaminant Hydrology* **7**: 51–73.

Lu, Z. G., H. M. Zhang & J. S. Wang. (1986): A satellite remote sensing and meteorological model for estimating wheat yield over a large area. *Acta Agriculturae Universitatis Pekinensis* **12**: 205–210.

Neve, S. de, G. Hofman & S. de Neve. (1998): N mineralization and nitrate leaching from vegetable crop residue under field conditions: a model evaluation. *Soil Biology and Biochemistry* **30**: 2067–2075.

Norcliffe, G. B. (1977): *Inferential Statistics for Geographers, An Introduction*. Hutchinson, London.

Scholefield, D., D. R. Lockyer, D. C. Whitehead, & K. C. Tyson. (1991): A model to predict transformations and losses of nitrogen in UK pastures grazed by beef cattle. *Plant and Soil* **132**: 165–177.

Semenov, M. A. & J. R. Porter. (1995): Climatic variability and the modelling of crop yields. *Agricultural and Forest Meteorology* **73**: 265–283.

Spector, P. E. (1981): *Research Designs*. Sage University Paper 23, Quantitive Applications in the Social Sciences. Sage Publications, CA.

Thomas, R. W. & R. J. Huggett. (1980): *Modelling in Geography, a Mathematical Approach*. Harper & Row.

Tyler, S. W. & S. W. Wheatcraft. (1990): The consequences of fractal scaling in heterogenous soils and porous media. In *Scaling in Soil Physics: Principles and Applications*. SSSA special monograph 25. Soil Science Society of Ammerica, Madison, WI., USA, p 109–122.

Soil Survey Staff (1951): *Soil Survey Manual*. United States Department of Agriculture Handbook No. 18. United States Department of Agriculture, Washington DC, USA.

Wilmot, C. J., S. G. Ackleson, R. E. Davis, J. J. Feddema, K. M. Klink, D. R. Legates, J. O'Donnell & C. M. Rowe. (1985): Statistics for the evaluation and comparison of models. *Journal of Geophysical Research* **90**: 8995–9005.

Chapter 2 Introduction to Geographic Information Systems, Remote Sensing and Geostatistics

J. Aherne[1], W. Vullings[2] and N. M. Holden[3]
1. Department of Environmental Resource Management
    University College Dublin, Ireland
2. Department of Crop Science, Horticulture and Forestry
    University College Dublin, Ireland
3. Department of Agricultural and Food Engineering
    University College Dublin, Ireland

Geographical information systems (GIS), remote sensing (RS) and geostatistics are three relatively new tools central to agro-environmental modelling. They are important because they create (RS), manage (GIS) and analyse (GIS, geostatistics) spatial data which is fundamental to many modern modelling applications. Each technology will be considered from a theoretical and applied point-of-view in this chapter.

2.1 A brief introduction to Geographic Information Systems

2.1.1 What is a Geographic Information System?

A Geographic Information System (GIS) is a computer-based system with the primary function of acquisition, storage, manipulation, analysis and display of geographic data. In other words a system for managing spatial data (Ronald Eastman, 1995). *Geographic* implies that locations of the data items are known in terms of geographic co-ordinates; *information* implies that the data in a GIS are organised to yield useful knowledge; and *system* implies that a GIS is made up from several interrelated and linked components with different functions, typically including: spatial and attribute database, cartographic display, map digitising, database management, geographic analysis, image processing and statistical analysis.

Central to a GIS is a database that is a collection of maps or data layers in digital format. The database is comprised of two elements: a spatial database describing a position on the earth's surface and an attribute database describing a feature at that point. For example we might have the co-ordinates for the top of Carrantuohill mountain, Co. Kerry in the spatial database and the rainfall, temperature, or land cover at that point in the attribute database. Essentially, GISs are computer-based systems that can deal with virtually any type of information about features that can be referenced by geographical location.

As pointed out by Legg (1992), "There is nothing fundamentally new about the concept of a GIS, it is just the scale and speed of data handling which has changed". GIS technology takes common database operations such as query and statistical analysis and integrates them with visual and geographic analysis benefits offered by maps. These abilities distinguish a GIS from other information systems and make GIS valuable to a wide range of applications in the academic and professional environment.

2.1.2   Why use a Geographic Information System?

The ultimate purpose of GIS is to provide support for making decisions and conducting research based on spatial data, through the ability to spatially interrelate multiple types of information stemming from a range of sources (Bonham-Carter, 1994). GIS achieve this through one or more of the following operations on spatial data: organisation by spatial location, visualisation, query, combination, analysis and prediction.

A geographical database is organised in a fashion similar to a collection of maps. However they differ from paper maps in that digital maps that were derived from paper maps of different scale but covering the same geographical area, may be combined. In short, GISs are designed to bring together spatial data from diverse sources into a unified database, and to organise them as a series of data layers, all of which are geo-referenced, i.e. they overlap correctly at all locations (Figure 2.1).

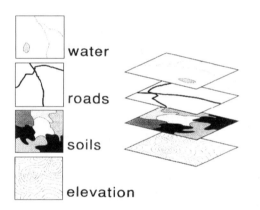

Figure 2.1: **In a GIS diverse data sources can be stored as data layers that are geo-referenced (Ronald Eastman, 1995)**

After visualisation the most fundamental of all the tools provided by a GIS are those involved with database query. GIS provide two types of interactive query: query by location, or query by attribute. The first type answer the question "What are the characteristics of this location?"; and the second type answer "Where abouts do these characteristics occur?".

The ability to manipulate and combine spatial data sets from different sources can often lead to an understanding and interpretation of spatial phenomena that are not apparent when individual spatial data types are considered in isolation (Bonham-Carter, 1994). Modelling in particular, requires that we be able to combine layers according to various mathematical combinations. For example, Goodale *et al.* (1998a) used an elevation map and polynomial regression within a GIS to model long-term average monthly precipitation, temperature and solar radiation changes with altitude for Ireland. This provided a very powerful method of spatially extrapolating data using the techniques described in Chapter 3, or as is possible in modern GIS, by using geostatistical methods. Chapter 6 illustrates how a number of data layers can be combined with (fairly) simple models to create a complex management system.

2.1.3   Data types

Not all Geographic Information Systems use the same logic for representing map data in digital form. However, most use one of two primary formats: a vector (polygon) or raster (grid cell) format. In *vector* systems the geographic representation of features are defined by a series of points that when joined with straight lines form the outlines of the features. The points are defined by a pair of numbers giving the X and Y co-ordinates. The attributes of the features are then stored with a traditional database management program. With *raster* systems the geographic location of features and the attributes they possess are merged into one data file. The features are not defined, but instead the represented area is subdivided into a fine mesh of grid cells in which the attribute of the point is recorded (Ronald Eastman, 1995). In this way each cell is given a numeric value which may represent a property like soil type, rainfall or temperature. The different data grids can be thought of as layers, each one of which stores one type of information over the mapped region. For example, for a mapped region we might have a map of elevation and a map of soil type, producing as many maps or data layers as there are information parameters (Figure 2.1).

Raster and vector data formats each have their advantages and disadvantages. Raster systems tend to have simpler data structures; they afford greater computational efficiency in such operations as overlay analysis; and they represent features having high spatial variability more effectively. On the other hand, raster systems are typically data intensive since they must record data at every cell location. However since geographical space is uniformly defined in a simple and predictable fashion, raster systems tend to be very rapid in the evaluation of problems that involve various mathematical combinations of the data in multiple layers (Ronald Eastman, 1995). Vector systems, unlike raster systems, tend to be more database management oriented. They are quite efficient in their storage of map data because they only store the boundaries of features and not what's inside the boundaries, and they excel at problems concerning movement over a network. The vector model is well-suited for representing maps. For many, it is the powerful database management functions and excellent mapping capabilities that make vector systems attractive. However, certain operations (e.g., overlay analysis) are more complex computationally in vector data format than in a raster format (Lillesand and Kiefer, 1994). Most modern GISs support both raster and vector data formats, taking advantage of the good characteristics of both.

2.1.4   Data acquisition

A major proportion of the effort in any GIS project involves assembling the data in digital form and creating a spatial database in which all the maps, images and spatial tables are properly geo-referenced (Bonham-Carter, 1994). Map projection is a necessary obstacle to be overcome before the interesting part of the study can begin. Co-ordinate conversions are often required, typically involving the transformation of table co-ordinates from a digitiser to the co-ordinates of a map projection, conversion of projection co-ordinate to geographic co-ordinates, and the conversion of arbitrary raster co-ordinates to the co-ordinates of some known projection. GIS packages provide utilities for changing the projection and reference system of digital layers. However, before data layers can be registered to the same co-ordinate system they have to be incorporated into a GIS. There are a number of methods of map acquisition such as, digitising, remote sensing and image processing (see Section 2.2), and surface mapping techniques.

The digital capture of data from maps is carried out in one of two principal ways: manual digitising and raster scanning using optical scanners. Manual digitising uses a digitising table which is equipped with a stylus or cursor for tracing and electronically recording the positions of points and lines. The map is mounted on the table, taped securely to prevent movement, and traced with the cursor. The position of the cursor is transmitted from the table to the computer where a digitising program processes and stores the raw data. The resulting strings of spatial co-ordinates comprise the raw data for the vector representation of points, lines and polygons. Scanners may also be used to digitise data such as aerial photographs. The result is a graphic image, rather than outlines of features such as are created with a digitising tablet. GIS can typically convert between vector and raster formats.

Surface mapping is a general term for the estimation of surface characteristics from irregularly-spaced point data. This point-to-area conversion can be carried out in numerous ways, depending on the level of measurement of the point attribute data. Surface modelling involves interpolation from irregularly-spaced samples to a grid of points. The interpolation process involves estimating the value of the modelled variable at a succession of point locations usually on a square lattice. The grid values are then treated as the pixels of a raster image. Alternatively, contour lines are threaded through the grid and the data are represented in a vector structure. There are many different approaches and methods to surface mapping (such as triangulation, distance weighting and kriging); for more information the reader can consult a text such as Isaaks and Srivastava (1989).

2.2   A brief introduction to Remote Sensing and Image Processing

2.2.1   What is Remote Sensing?

Remote Sensing (RS) has been defined by Lillesand and Kiefer (1994) as the science and art of obtaining information about an object, area, or phenomenon through the analysis of data acquired by a device that is not in contact with the object, area, or phenomenon under investigation. This definition, like most definitions of remote sensing, is ambiguous, as even reading this book is defined as "remote sensing". In this section we will present an overview of the concepts of remote sensing. Figure 2.2 indicates the stages and procedures involved in the application of remote sensing, each of which will be considered briefly. For more detailed information see Lillesand and Kiefer (1994).

Electromagnetic energy occurs in different forms. Visible light is of course the most known form, but heat, radio waves, X-rays and ultraviolet rays are also very familiar. All this energy is inherently similar and radiates in accordance with basic wave theory (Lillesand and Kiefer, 1994). When electromagnetic radiation reaches a surface, three specific interactions can occur. A part of the radiation can go through the object, a part can be reflected and a part can be absorbed (Molenaar, 1990). The proportion in which the radiation is divided over the three pathways depends on the material type or in remote sensing terms, 'objects have their own spectral signature' (see Figure 2.3). This response makes it possible to distinguish between different objects. In the visible wavelength range (0.4–0.7 µm) this phenomenon causes colour. For example, vegetation looks green because it absorbs most of the radiation in the red (0.6–0.7 µm) and blue (0.4–0.5 µm) wavelength range and reflects highly in the green (0.5–0.6 µm) portion of the spectrum, a phenomenon depicted in Figure 2.3.

Electromagnetic radiation can be measured by sensors. There are two different types of sensors available: passive and active. Passive sensors measure radiation that originates from natural sources, such as sunlight reflected by an object or energy emitted by an object. They do not emit radiation themselves. Active sensors, such as radar, have their own radiation source.

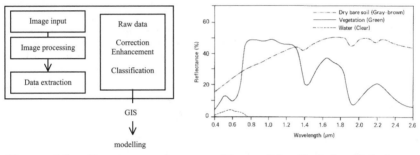

Figure 2.2: Remote sensing concepts; stages and methods used in the application of remote sensing

Figure 2.3: Typical spectral reflectance curves (signatures) for vegetation, soil and water, (Lillesand and Kiefer, 1994)

2.2.2   Image input

After the decision to use remote sensing, the next question will be: "what kind of data are needed?" Sensor systems can be raised from the ground, carried in aircraft or on satellites. Each system produces a different kind of raw data. Many factors influence the choice of data, not only factors determined by the kind of system used, but also, for instance, costs for purchasing images, and cloudiness where wavelengths of the visible spectrum are used.

Raw data can be divided into analogue (photographs) and digital (e.g. satellite images, video and digital photography), and can be described by spectral, spatial and temporal resolution. The spectral resolution describes which wavelength bands the sensor detects. Spatial resolution is determined by the dimensions of the ground area that is sensed at any instant in time (Lillesand and Kiefer, 1994). The revisit time to an area is called the temporal resolution.

Most satellites are polar orbiting, but not all. Meteosat, a satellite dedicated to meteorology, is geo-stationary which means that its orbiting time is synchronous to the rotation of the earth at the equator, and hence its view of the surface of the earth remains fixed.

2.2.3   Image processing

Before information can be extracted from raw data it must first be corrected for errors made during the actual sensing process. Factors that disturb the sensing process can be divided into radiometric distortions and geometric distortions. The corrections that are needed to reduce the distortions for a particular set of data depend on the type of distortions, the type of data and also on the expected use for the data. After the data have been restored, enhancement techniques can be used to emphasise certain aspects that will help extract the required information.

Factors such as atmospheric conditions, viewing geometry and instrument response characteristics cause radiometric distortions. The kind of correction necessary depends on the type of distortion but also on the application of the data. When comparing different satellite images of the same area in the visible and near infrared wavelengths, sun elevation correction, earth-sun distance correction, and haze correction, might be useful to overcome differences caused by the atmosphere. The distortion caused by sensing, signal digitisation or the data recording process, known as noise, typically causes problems like striping or banding. Most image processing (IP) software packages have built-in standard options for correction of these errors.

During the process of taking satellite images, factors such as earth rotation, earth curvature, sensor characteristics and satellite movement cause geometric distortions. Although these distortions often look less disturbing than the radiometric distortions, they are usually far more severe and have to be corrected before the images can be integrated with a GIS. There are two different techniques for correcting geometric distortions. The first approach tries to model the type and magnitude of the distortions and generate correction formulae; this is only possible when the types of distortions are well understood. The second one is based on statistics and tackles the problem without knowing the source and magnitude of the distortions. The exact transformation functions are not known, but polynomial functions (mostly of the first, second or third order) are used to represent them. This technique is the most frequently used procedure and can achieve within-pixel level precision (Wilkie and Finn, 1996). It is based on establishing an assumed mathematical relationship between the pixel locations in the remotely sensed data and the pixel locations in the master image or their co-ordinates on the ground. In both cases these points are called 'ground control points' and they have to be well defined and unambiguous, for instance, cross roads, buildings, bridges over rivers, etc. The polynomial functions are fitted to the control points using least squares regression. The order of the polynomial function that is used depends on the kind of geometric distortion, and the number of control points needed for a transformation depends on the order of the polynomial function. To solve a polynomial function of the first order, a minimum of three control points is needed, six for a second-order and ten for a third-order polynomial. In practise more points are used as the greater number make a better fit possible. The positional error in the procedure can be calculated and also improved by removing and adding control points. This procedure is known as 'rectification' and the end product is a geometrically corrected image (see also Sections 4.2.3, 4.2.4 and 4.2.5).

The goal of image enhancement is to improve the visual interpretability of an image by increasing the apparent distinction between the features in the scene (Lillesand and Kiefer, 1995). Image enhancement involves techniques such as contrast improvement, pseudo-colour assignment, noise smoothing and improvement of the sharpness of images. By these techniques certain aspects in the image can be clarified without pretending that the result shows a more reliable representation of reality (Rosenfeld and Kak, 1982). There are many different enhancement techniques available. The decision about which technique to use depends on the type of image (e.g. black/white, colour mode, radar, number of wavelength bands) and on the kind of output that is required. Enhancement techniques can be divided into three groups. The first group covers the techniques that manipulate contrast in satellite images (e.g. contrast stretching, grey-level thresholding, and density slicing). The second group includes methods that manipulate spatial features (e.g., edge enhancement, spatial filtering, and fourier analysis). The last series of techniques is only useful when more images or wavelength bands of the same area are available (e.g., ratio images, principal component analysis and colour space transformations).

2.2.4   Data extraction

The human eye is a perfect tool for interpreting satellite images. It is able to recognise colour, texture, shapes and patterns and it combines those elements when interpreting images. However, common automated classification techniques are solely based on spectral radiance (spectral pattern recognition). Procedures that include shapes, patterns and sizes (spatial pattern recognition) are available, but these techniques are not yet as reliable as the spectral pattern recognition techniques.

There are two different approaches within spectral pattern recognition: the unsupervised and the supervised classification. An unsupervised classification is preferred when there are distinct spectral radiance classes within the image and their exact 'meaning' (i.e. the physical property represented) is not required. When extraction of specific features is required a supervised classification is best. Figure 2.4 shows that when the digital numbers of two bands are plotted against each other a clustering occurs under ideal conditions. The utility of remote sensing depends on statistical separability of available spectra.

Unsupervised classification involves algorithms that examine the pixels in an image and aggregate them into a number of classes based on the natural groupings or clusters present in the image. The basic premise is that values within a given cover type should be close together in space and representing data in different classes should be comparatively well separated. The classes that result from unsupervised classification are spectral classes. Because they are based solely on the natural groupings in the image values, the identity of the spectral classes is not initially known. The analyst must

compare the classified data with some form of reference data to determine the identity
and informational value of the spectral classes (Lillesand and Kiefer, 1994).

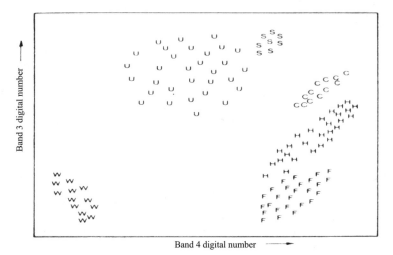

**Figure 2.4: Pixel observations plotted on scatter diagram (F: forest; H: hay; C:
corn; S: sand; U: urban and W: water (Lillesand and Kiefer, 1994)**

Supervised classification according to Lillesand and Kiefer (1994) typically involves
three basic steps: (1) the training stage where representative training areas are identified
and a numerical description of the spectral attributes of each land cover type of interest
is developed. (2) the classification stage where each pixel in the image data set is
categorised into the land cover class it most closely resembles, and (3) the output stage
where the entire categorised data set is presented as a new image. There are different
techniques to determine in which class a pixel belongs. Three techniques are commonly
used as classifiers: minimum distance to mean, parallelepiped and maximum likelihood.
Every classifier has advantages and disadvantages. They mainly differ in the time they
need for the analysis and the accuracy. For every application the analyst should judge
which classifier is the most suitable. Thus, in the supervised approach the analyst
defines useful information categories and then examines their spectral separability,
whereas in the unsupervised approach spectrally separable classes are determined
followed by definition of their informational utility (Lillesand and Kiefer, 1994).

In both classification methods ground-truth control becomes necessary at some stage.
The linking of the features in the image to real-world-features is the most important step
in the whole process. A Global Positioning System (GPS) can be used to help co-
ordinate between the image and ground positioning.

In the past years much research has been done in remote sensing and much is written
about linking real world data to images (Table 2.1). This information is very useful in
the classification and ground-truth processes but also in the earlier stage of determining

which kind of source (analogue, digital, or sensor system) will be helpful in solving the research problem and achieving the project objectives.

Table 2.1: Basic information available from remote sensing imagery

Biophysical Feature	Blue	Green	Red	IR Reflected	Thermal	Radar
x, y position	x	x	x	x	x	x
Shape, size, orientation	x	x	x	x	x	x
Topographic elevation	x	x	x	x	x	x
Bathymetric depth	x					
Subsurface location						x
Colour	x	x	x	x		
Temperature					x	
Texture/roughness	x	x	x	x	x	x
Vegetation:						
Chlorophyll abs.	x		x			
Biomass				x		
Moisture content				x		
Soil:						
Moisture content				x	x	x
Minerals:						
Hydroxyl ion content				x		
Fires					x	

Source: Wilkie and Finn, 1996.

2.3   A brief introduction to Geostatistical tools for spatial interpolation

2.3.1   Background Concepts

The data used in agro-meteorological modelling is almost exclusively point data. This means that each datum is associated with a given location on the earth's surface. If the location of the datum is known in a quantified manner, with respect to an understood co-ordinate system, then the datum can be said to be geo-referenced. It is common to have a range of data for a particular location (such as a weather station, or a soil profile). Simple distance weighted interpolation of data to provide values at a point where none exist is quite acceptable for many purposes. There are however, a number of statistical techniques available for analysing geo-referenced data with a view to developing a statistical model of the spatial patterns that exist. From the model it is then possible to predict values at locations between those sampled. To apply such techniques, as with all statistical approaches, requires that the data satisfy the assumptions of the underlying theory. In the next few pages, the basic techniques of spatial interpolation will be reviewed as they apply to both atmospheric and terrestrial variables. As this subject has been covered in a number of dedicated books and review papers (Clark, 1979; Isaaks and Srivastava, 1989; Rendu, 1981; Goovaerts, 1999; Oliver, 1999) this will be a general overview that will hopefully point the reader in the correct direction.

As a word of caution to the reader: it is necessary to point out that the material presented here is "the tip of the iceberg". The solution of complex problems should only be addressed once a solid understanding of the techniques and their assumptions has been attained. Application of analytical techniques without understanding their constraints will tend to result in meaningless misapplication. This is a view shared by many writers in the geostatistical area (see for examples Kitanidis, 1997), and cannot be over-emphasised.

Geo, to do with the earth, and *statistics*, methods for organising, analysing and presenting data, combine to produce the term *geostatistics*, the analysis of spatial data. From statistics arises *statistical modelling*, which is the fitting of equations to data in order to predict unknowns (also known as *empirical* modelling and, with the inclusion of probability can lead to *stochastic* modelling). Statistical modelling follows steps similar to any other modelling exercise. An equation is suggested that describes the variability in the data and can be used to predict unknowns (e.g. a regression equation: y $= a + b$x); the parameter values are estimated (a and b); the model is tested (significance of the parameters and the co-efficient of determination are evaluated); and if found to be suitable, the model is accepted and used. The more common equations found in empirical modelling such as linear, polynomial, exponential regression, work by reducing error (least-squares fit). Matheron (1963; 1971) adapted such methods to be more suitable to spatial problems.

Spatial interpolation is used to estimate a value of a variable based on other known values from surrounding sites. There are two common approaches (Kitanidis, 1997): (i) *point estimation* - which is the use of variables at a number of points to estimate the value of the variable at a given, non-measured point. (An example of this might be to predict the temperature at a given location in order to use a crop model), and (ii) *averaging* - here, point estimates of a value are used to establish an average value for an area. If the mean areal precipitation over a watershed that is well instrumented with a number of rainfall gauges is known, then to calculate a mean value the arithmetic mean could be used, or a more complex approach would be to investigate spatial pattern and apply some type of spatial weighting.

The interpolation of data results in *best estimates* of what the unknown value *might be*. The main advantage of a statistical approach to interpolation is that it also permits an estimation of *error*. The estimation of error allows us to decide how much faith to put in the interpolated data. A distance-weighted (linear or similar) approach provides no assistance in this regard.

2.3.2   Describing spatial patterns

The description of spatial patterns is useful in order to evaluate the nature of available data. If we wish to know something about a location without data, we need to interpolate for that point. Interpolation of irregular data into a grid of data is also a common first step in preparing contour maps of variables. For analysis of data in a GIS it is sometimes recommended that irregular data first be gridded. There are a number of approaches available and selection of the best can be assisted by knowledge of the spatial pattern of the original data available. The description of spatial patterns can have varying degrees of complexity. We will look at some of the available options.

Distribution of points in space
The distribution of points in space can be regarded as uniform if the density of point in any sub-area is equal to the density of points in all other sub-areas of the same size (Davis, 1986). The pattern is regular is the points are on a grid. The quality of an interpolation depends on the density, uniformity and regularity of the data. With low density data it is probably better to use simpler techniques.

To test for uniformity, the whole sample area is split into equal-sized sub-areas and the points per area are then statistically evaluated using a χ^2 test. The expected number of points per sub-area can be calculated as:

$$E = \frac{n}{T}$$

2.1

. where n = number of points and T = number of sub-areas. From this the χ^2 statistic can be calculated as:

$$\chi^2 = \sum \frac{(O-E)^2}{E}$$

2.2

where O = the number of points in a sub-area and there are T-2 degrees of freedom (Davis, 1986). If the size of the sub-area is changed then the result of the analysis can change. By evaluating a range of sub-area sizes it could be possible to identify optimum management areas for applying models based on the available data for a region.

It is also possible to test whether points are randomly located or are regular using a poisson distribution comparison because it is possible for both cases to be uniform. Similarly clustering of data can be evaluated by comparison with a negative binomial distribution (Ripley, 1981). The reader is directed towards Davis (1986) for more detail.

Nearest neighbour analysis is a viable alternative to using sub-areas to test distributions of points (Davis, 1986; Ripley, 1981). In this case, the distance between each point and its nearest neighbour is compared with those expected from a random distribution. For a given number of points (n) in an area (A), the mean distance between neighbours is:

$$\bar{\delta} = \frac{1}{2}\sqrt{\frac{A}{n}}$$

2.3

The variance of $\bar{\delta}$ is:

$$\sigma_{\bar{\delta}}^2 = \frac{(4-\pi)A}{4\pi n^2} = \frac{0.06831A}{n^2} \qquad 2.4$$

The standard error of the mean distance between neighbours is the square root of the variance:

$$S_e = \frac{0.26136}{\sqrt{\dfrac{A}{n^2}}} \qquad 2.5$$

If n > 6, then a z-test can be used to evaluate whether the observed mean distance between nearest neighbours ($\bar{d}$) is different from that of random points:

$$z = \frac{\bar{d} - \bar{\delta}}{S_e} \qquad 2.6$$

The value of $\bar{d}$ is calculated by finding nearest neighbours for each point and reporting the mean distance of separation. This procedure assumes that the edge effects of the area are not biasing the results. The nearest neighbour statistic (R) is then:

$$R = \frac{\bar{d}}{\bar{\delta}} \qquad 2.7$$

where the magnitude indicates the nature of the point arrangement as:

R =   0.00 - all points lie at the same co-ordinate
         1.00 - random locations
         2.15 - a perfectly spaced hexagonal grid.

These quantified evaluations of point patterns are often ignored in favour of by-eye judgement, but for a complete analysis of spatial pattern they should be considered.

The experimental semivariogram
An alternative to describing and analysing spatial point patterns is to evaluate the spatial dependence of the value at a particular location to the values at points around about. A datum with a known location is called a *regionalised variable*. The rate of change with distance of regionalised variables with distance is described by a semivariogram ($\gamma(\mathbf{h})$). The term variogram (strictly twice the semivariogram) is now used interchangeably with semivariogram which can cause some confusion for beginning researchers.

A semivariogram can be constructed for continuous or categorical variables. In the case of a continuous variable (z), measured at n locations $\mathbf{u}_\alpha$, the spacing of two variables can be described by the vector **h**, the distance between pairs of values. The experimental semivariogram is a representation of the average "dissimilarity" between pairs of data separated by **h**, where the x-axis represents the separation distance or *lag*, and the y-axis represents half the average squared difference between the paired data at the particular lag. Given a sample of observations $z(\mathbf{u}_\alpha)$, $\alpha = 1, 2, ..., n$ then:

$$\hat{\gamma}(\mathbf{h}) = \frac{1}{2N(\mathbf{h})} \sum_{\alpha=1}^{N(\mathbf{h})} [z(\mathbf{u}_\alpha) - z(\mathbf{u}_\alpha + \mathbf{h})]^2 \qquad 2.8$$

$\hat{\gamma}(\mathbf{h})$ is the semivariogram, which unlike the mean is not a single number, but a continuous function of the lag **h**. Being a vector, **h** includes both distance and angle between points (Figure 2.5). α is the sample number, with each sample being described by 2 co-ordinates and a magnitude. In a simple case of points on a line or transect, the lag is the inter-point distances. In calculating the experimental semivariogram pairs of data that have the same lag are averaged and it is assumed that the effect of location is removed leaving a dependency on distance. This is an assumption of the technique and there is no method to test whether this assumption is tenable. It is therefore also assumed that the mean of the regionalised variable is consistent over the whole area. If this is not the case there is systematic variation in the mean, and this effect is known as *drift*. Where drift is known to occur it can be removed using either a model of the drift (Olea, 1975) or filtering (Christakos, 1992). This situation complicates application of geostatistics to weather variables in Ireland where there are distinct systematic

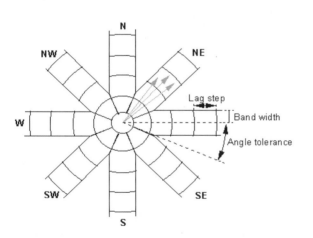

Figure 2.5: The concept of the lag, including distance and direction

variations due to altitude, coastal effects and land mass effects. A typical, spatially limited semivariogram (Figure 2.6) shows increasing dissimilarity over short lags and then levels off. The lag at which the curve levels off is known as the *range* (*a*) and the value of the semivariogram beyond the range is known as the *sill*. The *nugget* is the value of the semivariogram at a lag of zero. The experimental semivariogram obtained from sample data cannot be compared with a "true" semivariogram by any statistical method; rather, user judgement is called into play. Where there is no sill the semivariogram is not spatially constrained.

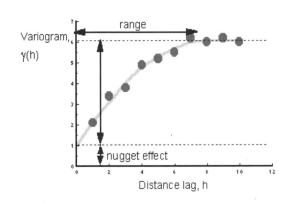

Figure 2.6: The main features of a spatially limited semivariogram model

The patterns of spatial dependence that characterise a regional variable are detected and modelled using *structural analysis* (Olea, 1975). This is the step of fitting a model to the experimental semivariogram. In order to do this, some sampling concerns have to be addressed. Analysis is facilitated when samples have been taken from equally spaced grids or transects because the pairing of data into lags is easier. A random strategy results in few paired data of the same lag while the grid strategy results in pairs based on directions parallel to the grid, and at various angles. If it is not possible to obtain many paired data for each lag, then a series of classes (0 - 1, 1-2, 2-..., -n) can be used (Isaaks and Srivastava, 1989). One feature of semivariograms is that the number of pairs used to calculate the semivariogram at a given lag will tend to decrease as the lag length increases, thus the accuracy of the estimate will tend to decrease with increasing lag as well. One rule-of-thumb that is used is to only calculate lags with a minimum of 30 pairs of data (c.f. Journal and Huijbregts, 1978). It will be immediately apparent from studying the details of the weather stations in Ireland that such a rule could not be applied to interpolation of weather data very easily. Another general recommendation is to make the maximum lag distance equal to half the maximum distance in the experimental field (Journal and Huijbregts, 1978).

A final consideration is that of *support*. Support is the term used to describe the size and shape of the sample taken to define the variable value at a particular "point". In measuring properties such as soil hydraulic conductivity, crop yield and even air temperature, the size of the sample (or the physical environment in which it is taken) will influence the value obtained. Thus when preparing data for geostatistical analysis it is essential that comparable samples are used - the hydraulic conductivity of a soil as characterised by rings 5 cm diameter will be very different from the same soil characterised by rings of 50 cm diameter. These issues can be addressed and the reader is advised to become familiar with the concept (see, for example, Clark, 1979; Rendu, 1981) prior to attempting any analysis.

The semivariogram model
In order to use the information contained within the experimental semivariogram, it is necessary to fit a model which is a function of the lag value **h**. Common models that are acceptable are the spherical, exponential, power and Gaussian semivariogram models (Olea, 1975; McBratney and Webster, 1986). Examples of two models are:

The *spherical semivariogram model*:

$$\gamma(h) = \left| \begin{array}{c} C\left[\dfrac{3h}{2a} - \dfrac{1}{2}\left(\dfrac{h}{a}\right)^3\right] \quad \textit{where} \quad 0 \le h < a \\[2mm] C \qquad\qquad\qquad \textit{where} \quad a \le h \end{array} \right. \qquad 2.9$$

The *exponential semivariogram model*:

$$\gamma(h) = C\left(1 - e^{-\frac{3h}{a}}\right) \qquad 2.10$$

where C is the sill value (in measurement units squared), and a is the range (distance units). For the exponential model, a can generally be taken to be equal to $0.95C$.

It is possible to make far more complex models by linear combination of any permissible model with positive coefficients. This is referred to as a *nested semivariogram model*. If a single model does not fit the complexity of the structure observed in the experimental semivariogram then a nested model is necessary. Bogaert *et al.* (1995) offer seven models for use with meteorological data, and the reader should consult their documentation for details.

Having established that there is a large number of models available and that these can also be nested, how is the model to fit to the experimental semivariogram selected? Goovaerts (1997) indicates that the choosing of a model, and the estimation of its parameters is a controversial issue. Methods range from "black-box" procedures with full automation, to fitting a model by eye until it is graphically acceptable. The middle ground is to choose the type of model by eye, and to then find parameter values by automatic least-squares methods. Goovaerts (1999) is of the opinion that black-box methods should be avoided because they cannot make use of additional information known to the researcher. He is also wary of too much reliance of statistical curve fitting because the best model to capture the essence of the spatial pattern may not be the best model by statistical criteria. The experimental semivariogram is only one element of the information available and should not be viewed as being the exclusive guide. In general, the simplest model for the job is the best.

The procedure to follow is: (1) choose the type of model; (2) fit the model by trial and error using computer software; (3) evaluate the model for anisotropy and drift. The semivariogram should be analysed for at least four directions. If there is no difference with angle, then the semivariogram is isotropic.

There are no significance tests for geostatistics to assist in the evaluation of a semivariogram model; however *cross-validation* permits the researcher to evaluate the results in a manner that is common, if not widely used, wherever there is a lack of data.

2.3.3   Spatial interpolation

The intention of this section is to introduce the reader to geostatistical interpolation of data. Specific techniques relate to meteorological variables that are presented in Chapter 3, but most of the following discussion is relevant to any agro-environmental data. The most important use of geostatistics in the context of modelling is the interpolation of data to non-sampled sites. We will look at some approaches to spatial prediction as a guide to techniques that are available. As ever, the proviso exists that you should try to understand what each method does prior to application using automated software in order to detect artefacts and poor results. There are a number of approaches that can be taken; the following are selected examples.

Nearby value
The simplest method of interpolation is to adopt a nearby value that is known and apply it to the unknown site. In the case of weather data it is common enough to adopt a value from a nearby point of observation (such as a synoptic station) when it is thought that the variable in question does not change greatly over a wide area. A rigorous approach to applying this method is to construct *thiessen polygons* around each known data site and to allocate nearby values on the basis of where the target site lies. A thiessan polygon is an irregualr shape that contains all points that are closer to its known datum than to any of the alternatives. The algorithms for calculating thiessian polygons are presented by Davis (1986).

Distance weighting
Gridding is an important pre-treatment for data prior to analysis using GIS tools. The tools used to create a grid from irregular data are also useable for interpolation to any unknown location, provided it has known data points surrounding it. The most common method of interpolation is the distance weighted average. Rather than taking the average of surrounding data values, the points used are weighted based on their distance from the target point. One consequence of this method is that interpolated values *always* lie between the minimum and maximum observed values. The distance of each known point to be used in the averaging is calculated using Pythagorus:

$$D = \sqrt{(x_t - x_k)^2 + (y_t - y_k)^2} \qquad 2.11$$

where D = distance, x and y are the co-ordinates, t = target site, k = known site. The value at the target site is then estimated using the distances of the known sites as weights:

$$T = \frac{\sum_{i=1}^{n}\left(\dfrac{x_i}{D_i}\right)}{\sum_{i=1}^{n}\left(\dfrac{1}{D_i}\right)} \qquad 2.12$$

where T = value of variable at target site, x = value of variable at known site i. When using a dense dataset for interpolation it is normal to restrict such calculations to only the nearest points, or under some conditions, using points evenly distributed around the target, rather than just within a certain distance.

Kriging
Kriging is a generic term used to describe a family of least-squares regression algorithms used for spatial interpolation. These methods are significantly more complex than the simple interpolators presented previously. There are many methods (simple, ordinary, universal, cokriging, kriging with drift...), the mathematical details of which can be found in Clark (1979), Isaaks and Srivastava (1989), Kitanidis (1997) and Goovaerts (1997). Kriging is a form of weighted averaging, whereby samples near the unknown site have more influence on the prediction than samples distant from the site. If the sum of the weightings is equal to 1, and the prediction is a linear equation with minimum variance, then the estimator is known as a BLUE (*B*est *L*inear *U*nbiased *E*stimator). What this means, in practise, is that the values predicted at locations with known values (the sample sites) will always be correct, with a variance of zero (when using punctual kriging, but not for block kriging). Kriging is a way of finding this best estimator.

The fundamental equation for kriging is a linear regression (Goovaerts, 1999):

$$Z^*(u) - m(u) = \sum_{\alpha=1}^{n(u)} \lambda_\alpha(u) [Z(u_\alpha) - m(u_\alpha)]$$ 2.13

where $Z^*(\mathbf{u})$ is the linear regression estimator (the predicted value of the variable (z) at position $\mathbf{u}$), $\lambda_\alpha(\mathbf{u})$ is a weighting assigned to the datum $z(\mathbf{u}_\alpha)$, which is a realisation of the variable $Z(\mathbf{u}_\alpha)$, found within a given neighbourhood, $W(\mathbf{u})$ and $n(\mathbf{u})$ are weights chosen to minimise the estimation or error variance and are obtained by solving a system of linear equations (the details of which are not necessary here) known as the "kriging system". $m(\mathbf{u})$ is a trend term. The weighting of each datum $(z(\mathbf{u}_\alpha))$ will reflect their position relative to that being estimated; those further away and known not to be so important will have much less weight than those close by and known to be spatially related. The semivariogram model is used when kriging, thus the model chosen to fit the experimental semivariogram is very important when it comes to spatial prediction.

In some cases, data available for interpolation are sparse or have poor spatial structure. The former situation can be the case with meteorological variables due to the density of measurement stations. Under such conditions it is helpful to use secondary information which is available at each sampling location *and* at the locations for which the variable of interest is to be predicted (known as exhaustively sampled). In this situation three further techniques are available: (1) kriging within strata, where the secondary information is used to stratify the primary information, and separate semivariogram models are developed; (2) simple kriging with varying local means, where each strata is used to define a distinct local mean value; and (3) kriging with an external drift, which is similar to universal kriging, but rather than modelling the trend as a function of co-ordinate, it is modelled as a function of the secondary variable. In this case the relating function must be linear and make physical sense.

On occasions the secondary information available is not exhaustive, i.e. it is either not available at all sample sites, or for all sites to be predicted. Under such conditions, an alternative approach called *cokriging* is available. This is a multivariate extention of kriging.

Cokriging is much more demanding of data because more semivariogram information has to be obtained, related and modelled. In general, cokriging is only practical when the primary data are undersampled with respect to the secondary data, and when a strong correlation exists between the two (Journal and Huijbregts, 1978; Goovaerts, 1998). Goovaerts (1999) points out that another use of cokriging is where two methods of measuring the same variable are used, such as determining soil pH from samples in the laboratory (relatively slow and costly, but accurate) and using paper colour test strips (cheap but not so good). It would be possible to use the accurate method to improve predictions based on the cheap method.

Observations on the application of spatial interpolation
There are now a large number of freeware, shareware and commercial geostatistical software packages available. Assuming that these have all been programmed correctly, the onus is on the user to make some apparently simple decisions: (1) is an automated package required, or is an element of user trial-and-error better? (2) is a very flexible package needed, or would a simpler, limited package mean a smoother journey through the minefield of analysis? (3) would it be helpful to have software that can prompt the user with analysis information to aid in model selection? (4) is it worth investing in expensive software? (5) will the package take data in the format collected, or will there be significant work in pre-processing the data for analysis? It is not necessary to fully understand the underlying equations used in geostatistics, because of the proliferation of software available for analysis. This means anyone can start working on the interpolation of data. It cannot be emphasised enough however, that an understanding of the constraints of each method is necessary.

2.4   Case study 1: Use of remotely sensed data with GIS

2.4.1   Problem definition

The mangrove forests in the coastal areas of Vietnam have suffered severe damage over the past 50 years. Population growth and the rapidly expanding aquaculture in these areas have put pressure on the forest resources. The general objective of this research was to produce an information database describing present land use for the salt water intruded part of the Mekong delta, with a view to the development of a land-use plan based on integration of land use, potential and physical conditions (e.g. weather). In this research GIS and remote sensing were used to aid the inventory process. GIS was mainly used to store the information and to present it in an accessible and well-organised manner. Satellite remote sensing was instrumental in quickly producing high quality land-use inventories and soil surveys. The case study illustrates the basic requirements for defining 'land' elements of any spatially variable agro-climatic model based on a GIS.

In this research only the reflected part of the visible and near-infrared light emitted by the sun, which was measured by a passive sensor was considered. SPOT HRV satellite images of different areas and years were used. The images were made in the colour-infrared mode. This means that the green wavelength was represented by a blue colour in the image, the blue wavelength by a green colour and the infrared wavelength by a red colour. The ground resolution of this mode was 20 metres.

2.4.2   Data Sources

According to Ehlers (1997) four predominant data sources must be considered in an integrated GIS: cartographic, socio-economic, field and image. The data used in this research can be classified as:

1. Cartographic data: Converted into digital format from maps that follow cartographic rules and specifications.
   - Soil map of the Mekong Delta, scale 1:25 000.
   - Inundation map of the Mekong Delta.
   - Topographic information (villages, hamlets, canals and roads).
   - Elevation map derived from aerial photographs using a stereoscope.

   Since no meteorological measurements were taken in the study area, the data were acquired from the nearest measurement station and used as constant values for the 10 by 10 km study area. A more detailed analysis using interpolations would only have been possible where suitable relationships could be established. (See Chapter 3 for detail of generally available meteorological data, and Chapter 4 for detail of other environmental data available and specific cartographic considerations for its use).

2. Socio-economic data: These non-spatial data can only be input into a GIS with additional locational information (e.g., co-ordinates). As socio-economic data are often aggregated, their geometric precision is usually rather low. Information was gathered about inhabitant numbers and facilities per hamlet.

3. Field data: Ground-truth data have always played an important role for validating accuracy in land-use classification and mapping applications of remotely sensed imagery. These data are increasingly being collected in digital formats and with GPS-controlled co-ordinate measurements. This allows for easy integration into GIS. Ground-truth research was carried out to find out what the different colours on the satellite images represented. In a few places in the study area the soil type was checked against the current soil map.

4. Image data: Images from airborne and space-borne platforms can be obtained in analogue or digital formats. Digital images are often seen as an inherent component of a raster-based GIS. SPOT HVR satellite images of 20 meter resolution were made in the colour infrared mode. Aerial photographs of the Mekong Delta were taken in 1992.

2.4.3   Processing satellite images and cartographic maps

The purchased images were of good quality and no further radiometric or geometric corrections were necessary for this research. The images were already geometrically corrected so the next step was image enhancement. The enhancement technique used for all images was "contrast stretching" (Figure 2.7). Figure 2.7a gives the original histogram and 2.7b shows what happens if no contrast stretching is applied; in that case only the display levels from 60 to 158 are being used. The simplest form of contrast stretching is linear stretch (Figure 2.7c). The smallest image value is assigned to the display level 1 and the highest image value to display level 255, so all the display levels are being used. The histogram stretch (Figure 2.7d), works according to the same principle as the linear stretching, but it also accounts for the frequency of occurrence of the image values. For special analysis, some features may need more enhancement and others can be completely left out. It is possible to assign all the display levels to a particular range of image values (Figure 2.7e). Of course, the first thing is to find out which part of the spectral range represents the feature of interest. When this information is known unnecessary spectral parts can be masked out and all the display levels can be assigned to the spectral part that is important. Image enhancement was undertaken to aid the user and potentially improve classifications.

On the basis of an unsupervised classification procedure the three major land use systems or cover types were distinguished; rice fields (bare soil on the satellite image because of the dry season), aquaculture and mangrove forest. All three areas needed more detail for further research, so further contrast enhancement was needed. To produce an image with more detail in the classified areas, such as aquaculture, a mask was laid over the spectral classes that represented mangrove forest and bare soil and all the display levels were assigned to the remaining spectral class, which represented aquaculture.

In order to incorporate the cartographic maps into the GIS they were first translated into digital format. The maps were digitised and the information on the maps was stored in attribute files. All the maps were stored in vector format, and depending on the kind of map, polygons, lines or points were used. Using the GIS made it possible to include more information, not only the obvious information, but also extra data were added to the same polygons, points or lines. For instance, the soil map was digitised and all soil units had attributes of the USDA classification system, the Vietnamese classification system, salt content and acid sulphate (pyrite or jarosite) associated with each polygon.

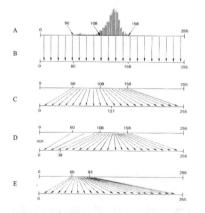

Figure 2.7: Contrast stretching (Lillesand and Kiefer, 1994)

Before inclusion into the GIS database the maps needed to be 'rectified', i.e., registered to the same co-ordinate space as the other maps. In this instance the satellite images were regarded as the master-images and control points that were visible and unambiguous on both the maps and the satellite images were needed to carry out the transformation. A fundamental requirement for integrated processing of remotely sensed data and GIS data is that they should be spatially referenced. Only then can we be assured of not trying to compare 'apples and oranges' (Ehlers, 1997).

2.4.4   Data extraction

The more recently available SPOT satellite image of the study area made during the dry season was used, which meant that the rice fields were dry. The satellite image indicated bare soil with different colours observed in the fields. The main objective of the field studies was to find out why the colour differences occurred and what the different colours represented.

Elevation was the primary factor that caused colour differences, but by an indirect mechanism. At the sites that had a higher elevation, the soils were more developed, they were drier and had better developed structure. These soils were covered with grasses, which resulted in a pink colour on the satellite image. The less developed and more saline soils were situated at lower elevations and had a pinkish white colour. In the few ploughed rice fields, the increase in surface roughness caused a decrease in reflection and this resulted in a dark green colour. The surface depressions were bright yellow, due to high reflection caused by an orange surface crust that formed during drying. This information allowed a classification of the SPOT image to be carried out.

Training areas of the different classes were selected and it was decided that the maximum likelihood classifier was best, because it determined the true shape of the distribution in each class and as a result was considered the most accurate (Wilkie and Finn, 1996). During the classification stage all the grid cells with a pink colour were put in class 1 (high elevation and more developed soils), all the grid cells with a pinkish white colour were put in class 2 (low elevation and less developed soils), the green grid cells were put in class 3 (ploughed fields), the cells with a blue colour were put in class 4 (moderate elevation) and the yellow grid cells were put in class 5 (depression, orange crust at surface). Bright red colour caused by trees (grid cells with this colour) formed class 6. The remaining grid cells could not be classified and formed the class 'unclassified'. This procedure resulted in a map with 7 classes.

2.4.5   GIS operations

After acquiring all the data in digital format it was easy to carry out overlay operations to compare maps. Validation of the soil augerings against the soil map was carried out using a point-area overlay. The co-ordinates of the soil augerings were acquired with a GPS and the soil information was stored as point information in the GIS system. This new point map was combined with the soil map, so the soil information per auger point could be compared to the information per soil unit of the map. Also the percentage of auger points of which the soil type matched the soil type given by the map could be derived. With this information it was possible to draw conclusions about the quality of the existing soil map. The higher this percentage the more accurate the map. It was also a good way of checking the differences between the Vietnamese and European soil mapping procedures.

One of the research questions was whether it would be possible to find a correlation between the salt content and soil acidity, and the relative height and position of the terrain. To answer this an area-area overlay of the soil map with the elevation map was carried out. The resulting map showed that the saline soils corresponded with the lower elevated areas and that signs of acidity were present only in the depression areas. It was concluded that there was a correlation between soil and elevation. Although this correlation was already noticed in the field, this method made it possible to quantify the areas.

2.4.6   Conclusions

The use of GIS and RS in this survey made the research possible in a short time period. Remote sensing images, and information extracted from such images, have become primary data sources for modern GISs. Indeed, the boundaries between remote sensing and GIS technology have become blurred and these combined fields will continue to revolutionise the inventory, monitoring, and management of natural resources on a day-to-day basis. Likewise, these technologies are assisting us in modelling and understanding biophysical processes at all scales of inquiry. They are also permitting us the development and communication of cause-and-effect 'what if' scenarios in a spatial context in ways never before possible (Lillesand and Kiefer, 1994).

The example presented indicates the basic requirements of any spatially sensitive agro-meteorological modelling in building up land information: (i) obtain spatial information; (ii) present in a spatial domain; (iii) process and integrate for maximum utility; (iv) ground truth; (v) analyse and classify. At this stage basic 'what if' scenario testing can begin, the physical environment has been modelled to a limited degree and interpretation can begin.

2.5   Case study 2: Application of a simple mass balance model through GIS[†]

2.5.1   Problem definition and hypothesis

One of the most pressing environmental issues over the last decade has been the need to predict the effects of acid deposition on terrestrial and aquatic ecosystems. As is typical of issues concerning public policy, a great deal of controversy surrounded the discussion of both the nature and extent of acid deposition effects. Ireland, being situated on the north western seaboard of Continental Europe, receives a relatively low level of anthropogenic acidifying substances via atmospheric deposition compared to the rest of Europe. However, recent Irish work has clearly shown that anthropogenic atmospheric deposition of measurable amounts occurs (Jordan, 1997). The loads measured have the potential to acidify poorly-buffered surface waters (Bowman, 1991) and to accelerate acidification in forest ecosystems (Farrell et al., 1993). While the concept of forest decline (reduced growth, needle discoloration) has become widely accepted, a recent

[†]   The research presented in the case study was funded by the Environmental Protection Agency under the Environmental Monitoring R&D Sub-programme of the Operational Programme for Environmental Services

assessment of forest health has argued against there being evidence of any new, widespread decline of European forests (Schlaepfer, 1993). Nevertheless, there have been widespread changes in the chemical status of forest soils over the past 100 years, and there are very real threats of a deterioration in forest condition as a result of climate change (Schlaepfer, 1993).

In considering the relative impacts of atmospheric inputs to the acidification of forest soils all of the potential sources and sinks of acidity must be taken into account. An increase in acid deposition load leads to an increase in the acidity of the soil. If the loss of alkalinity (neutralising capacity) by the soil is continually replenished by the dissolution of minerals from the soil parent material then the consequences are unlikely to be serious. However, if the rate of weathering is insufficient to neutralise the input of acidity, then there will be a reduction in the percentage of exchangeable bases and a parallel reduction in pH of the soil solution. In the long-term, soil buffering is dependent on the replenishment of exchangeable base cations from weathering of soil minerals. Therefore, the rate of chemical weathering within the soil and soil parent material largely determine its sensitivity to acidification. One method of quantitatively estimating the sensitivity of forest soils to acidification is to use the critical load approach (Sverdrup *et al.*, 1990). The critical load concept is strongly linked to sustainable management, since a deposition above critical load is not, by definition, sustainable in the long-term. The aim of the case study reported here was to quantify the sensitivity of Irish forest soils to acidification i.e., determine the ability of Irish forest soils to neutralise acidity, based on a critical load approach. This case study illustrates the data requirements and manipulations necessary to implement a "static" agro-climate model on a national scale.

2.5.2   Definitions and assumptions

The critical load concept has its origins in the "Target Loading" concept proposed by Canadian scientists during negotiations with the USA on transboundary pollution control in the early 1980s (Longhurst, 1991). The concept was further developed in Scandinavia, where the term "Critical Load" was used (Nilsson and Grennfelt, 1988). Critical loads differ from target loads in that the critical load is an inherent property of an ecosystem, while the target load is usually based on political decisions and can be set lower or higher than the critical load, for safety reasons in the first instance or economic reasons in the second instance (Nilsson and Grennfelt, 1988). The Critical Load concept has become a well-established part of the work programme of the United Nations Economic Commission for Europe (UNECE) on Long-Range Transboundary Air Pollution. During the course of 1989, 34 European nations agreed to determine and map critical loads, and with this knowledge to negotiate a protocol for reduction of sulphur emissions. These estimates of critical loads will be used by the Working Group on Abatement Strategies (WGAS) and the Task Force on Integrated Assessment Modelling (TFIAM) to optimise the abatement strategies for the whole of Europe (Hettelingh *et al.*, 1991). The approach focuses on the chemical changes in the environment and establishes these as the link between emissions and the biological environment. In principle it indicates a level of pollution beyond which an ecosystem is not sustainable. The critical load of acid deposition to soils has been defined as, "The highest deposition of acidifying compounds that will not cause chemical changes in soil leading to long-term harmful effects on ecosystem structure and function" (Nilsson and Grennfelt, 1988). A number of methods for the determination of critical loads are available (Hettelingh *et al.*, 1991). The most common methods are based on the application of mass balance equations, which calculate the balance between sources and sinks of acidity and set the critical load to the pollutant input at which a specific critical chemical value is not exceeded. The linking of ecosystem response to deposition level is central to the critical load calculation. In order to apply the concept four basic elements need to be defined:

1. Receptor: The ecosystem considered, such as forests or freshwaters.
2. Biological indicator: The organism selected to represent the receptor, such as the forest stand.
3. Chemical criterion: The chemical measure affected by the acid deposition that is used to predict the risk of damage to the biological indicator, such as the pH, concentration of aluminium, or the ratio of base cations to aluminium.
4. Critical limit: The most unfavourable value that the chemical criterion may attain without long-term harmful effects on ecosystem structure and function.

For each receptor ecosystem, a biological indicator is chosen. A suitable chemical criterion is selected for that biological indicator and a critical chemical limit is assigned. The critical limit is entered into a chemical mass balance equation together with all sources and sinks of acidity in the system. In this way, ecosystem response is linked to atmospheric inputs.

2.5.3   Model formulation

Soil acidification is defined as an increase in total adsorbed acidity in the soil in relation to the content of adsorbed base cations in the ion exchange positions and is generally interpreted as a decrease in base saturation (Sverdrup *et al.*, 1990). Initially in the acidification process a soil may acidify as it adsorbs hydrogen ions, in exchange for base cations, from acid percolate. Due to this *in situ* neutralisation the percolating solution may become neutral. Later, once storage of base cations has been exhausted, and the neutralising capacity with it, the acid percolate will no longer be neutralised. This is enhanced by the uptake of base cations by trees. The weathering of the mineral matrix will be the major long-term source of alkalinity to neutralise acidity in the system, as well as the major supply of base cations to replace removed base cations. The basic principle of the mass balance method is to identify the long-term average sources of acidity and alkalinity in the system and to determine the maximum acid input that will balance the system at a biologically safe limit.

All the sources of acidity in the system can then be put into a balance against all sources of alkalinity:

Alkalinity   Acidity

BC weathering + BC deposition = acid deposition + BC uptake + BC leaching   2.14

Where BC is non-marine base cations (calcium, magnesium, potassium and sodium). Rearranging Equation 2.14 so that acid deposition is on the left-hand side of the equation yields:

Acid deposition = BC weathering + BC deposition − BC uptake − BC leaching   2.15

By setting BC leaching to the maximum allowable i.e. the critical limit, allows the estimation of the maximum allowable deposition or the critical load.

Critical load = BC weathering + BC deposition − BC uptake − BC leaching (limit)   2.16

In the mass balance model, it is assumed that ion exchange is at steady state, and that there is no net change in base saturation. Nitrogen is ignored. Sulphur is assumed to be in steady state in the soil, that is no oxidation, reduction or uptake of sulphur. A simple hydrology is assumed, where there is only vertical infiltration through the profile. Weathering is assumed to be evenly distributed over the profile. The model assumes the soil solution to be at equilibrium at all times. Steady-state methods for calculating critical load imply that only the final result of a certain deposition level is considered. The time to reach this final stage is not considered important, but rather the fact that it will be reached sooner or later.

2.5.4   Stages in GIS application

As suggested by Bonham-Carter (1994), most GIS research projects have three major steps or stages: (1) bring all the appropriate data together into a GIS database and register them to the same co-ordinate system; (2) manipulate the data to extract and derive the spatial data layers relevant to the aims of the project (which in this case are the data layers for the estimation of sources and sinks of acidity); (3) combine the data layers. In this study all data layers were registered to the Irish National Grid System (Ordnance Survey of Ireland, 1996). A raster based GIS was used because it facilitates greater computational efficiency in operations involving overlay analysis. The procedure was:

Step 1:   The initial database building step was the most time-consuming phase. In this study it involved assembling the various required spatial data in digital format, properly registered so that each data layer overlapped correctly. The 1:100,00 CORINE land cover project (Ireland) (Ordnance Survey of Ireland, 1993) was acquired in digital format (vector) and translated into raster format. The 1:575,000 General Soil Map of Ireland (Gardiner and Radford, 1980) was digitised from the paper map and also translated into raster format. Long-term annual average rainfall and evapotranspiration data (1951–1980) were obtained from Met Éireann, The Irish Meteorological Service, (Fitzgerald, 1984). Long-term annual average rainfall chemistry data (1985–1994) were obtained from a number of monitoring stations (Jordan, 1997; McGettigan and O'Donnell, 1995; Farrell et al., 1993; Bowman, 1991). These data sets were entered into the computer to produce tables of easting, northing and parameter (rainfall, evapotranspiration and rainfall chemistry).

Step 2:   The second step involved processing the input layers to extract and derive the data layers relevant to the development of a forest sensitivity map. The CORINE dataset was reclassified to display only coniferous forests; the legend of the General Soil Map was recast into five weathering classes based on the Skokloster classification (Nilsson and Grennfelt, 1988). Similarly, three surface runoff classes based on data from Sherwood (1992) were established. Maps were generated for evapotranspiration, rainfall and rainfall chemistry using a surface mapping interpolation procedure to transform the point data tables into national coverage maps.

Step 3:   The third step involved combining the various data layers. This was carried out in stages with the ultimate map being the sensitivity of forest soils to acidification. The forest area map was transformed into a base cation uptake map by assigning tree uptake rates. Weathering classes were assigned base cation weathering rates. Limiting base cation leaching was derived by subtracting evapotranspiration and surface runoff from rainfall and then multiplying by the critical chemical limit (Section 2.5.3). Base cation deposition was derived by combining (multiplying) base cation concentration and rainfall. These four resultant maps were then combined, according to Equation 2.16, to produce the forest soil sensitivity map.

2.5.5   Conclusions

A description of the process (analogous to the conceptual model, see Chapter 1 for a definition) was an essential guide to the choice of data, the kinds of information extracted from it and the assignment of weights to establish a forest soil sensitivity map. The flowchart, or conceptual model, is a formulisation of the procedures (such as, data extraction, reclassification, overlay, arithmetic operations and surface mapping) required to produce the final map. In summary, analysis and modelling of spatial data in a GIS is not simply a matter of throwing the data layers into a "black box" computer program. A conceptual model (3 stage process as described in Section 2.5.4), preferably formulated in the early stages of a study, is used to guide various stages of GIS processing (Bonham-Carter, 1994). The example presented illustrates clearly how GIS permitted the essential spatial aspects of the problem to be addressed with a relatively simple point model.

2.6   Case study 3: A worked example of spatial interpolation

This example is not based on real data. It is a simple illustration of spatial interpolation using an illustrative dataset. The data are x and y co-ordinates (Figure 2.8) and values of temperature in °C (Table 2.2). The data will be analysed in sequence following the procedures outlined earlier in the chapter.

Uniformity?

The first question to address is: are the data uniform throughout the sample area? If we split the area into 4 squares, each 5 x 5, then the expected samples per square would be

17/4 = 4.25. The observed values are: top left - 4, bottom left - 5, top right - 4, bottom right - 4. Thus the χ^2 statistic is calculated as:

$$[(4 - 4.25)^2 + (5-0.25)^2 + (4 - 4.25)^2 + (4 - 4.25)^2] / 4.25 = 0.176 \qquad 2.17$$

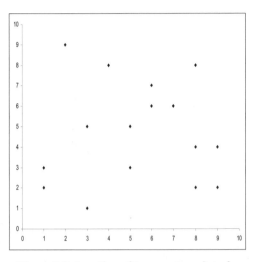

Figure 2.8: Location of temperature data for simulated example of spatial interpolation of data

The critical value of $\chi^2 = 5.99$ (taken from statistical tables), therefore we can accept the null hypothesis that there is no significant difference between the actual distribution of points and the expected distribution of points. The data points can be regarded as uniformly distributed given a grid size of 5 units. If the calculation is repeated using 25 squares of 2 unit side length, the result still indicates a uniform distribution of data

Randomness of distribution?
Having evaluated whether the points are uniformly spread, we can ask the question, are they randomly distributed, or do they approach a grid layout? We can evaluate this concern by considering at the z-value based on nearest neighbour distances.

Table 2.2: Data for illustration of geostatistical interpolation techniques

X co-ordinate	Y co-ordinate	Temperature (^{0}C)
1	2	9
3	1	12
8	2	17
5	3	16
9	4	15
6	6	12
3	5	10
2	9	9
7	6	14
6	7	14
8	4	16
5	5	11
9	2	16
1	3	10
4	1	13
4	8	11
8	8	14

The mean neighbour distance ($\overline{\delta}$) for a random distribution is:

$$0.5 \times \sqrt{(100/17)} = 1.213 \qquad\qquad 2.18$$

because the sample area (A) is 100 units and the number of points (n) is 17. The standard error (S_e) is therefore:

$$0.26136/\sqrt{(100/289)} = 0.443 \qquad\qquad 2.19$$

The mean distance between neighbours ($\overline{d}$) is calculated by finding the distance between each point and all the others, finding the smallest inter-point distance for each point, and then calculating the mean for all the points. For the small number of points in this example the calculation was performed in a spreadsheet by first identifying the point nearest to each one (from Figure 2.8), calculating the inter-point distance and then finding the average. The resulting value for $\overline{d}$ was 1.395. The z-test is calculated by:

$$(1.395-1.213)/0.443 = 0.411 \qquad\qquad 2.20$$

This result suggest that the points are not fully random but tend to be slightly clustered towards one spot in the area. If the data are reconsidered with respect to edge effects the result still stands. The simplest approach to evaluating the edge effect would be to treat the top edge as a continuation of the bottom edge, and the left edge as a continuation of the right edge. This exercise has no effect on the result. The result probably reflects the inherent bias of using an artificial data set where there is a tendency not to locate data at the edge of the area, thus causing a cluster towards a single point.

Interpolation
A number of approaches to interpolation are possible. We will look at each of them in turn. To permit comparison between the options, a simple task will be set as the example. The question being ask is, what would the temperature be at location 6,4 in the sample area from Figure 2.8.

Using a value from the nearest known point provides two possible results, site 4 (5,3) has a temperature of 16^0C and site 12 (5,5) has a temperature of 11^0C. To choose between these a researcher would have to use local knowledge.

The next logical step is to use the mean of the two nearest sites. In this case that is the same as a distance weighted interpolation using just 2 points, because both have the same distance. The temperature at the target site would be estimated as 13.5^0C.

A distance weighted interpolation using the 5 nearest sites would use sites 4(5,3), 6(6,6), 9(7,6), 11(8,4) and 12(5,5). The following calculations illustrate long hand, how to apply the distance weighted formula:

(calculation of distances)

Site	Co-ordinates	Temperature	Calculation	Distance
4	5,3	16	$\sqrt{(6-5)^2+(4-3)^2}$	1.414
6	6,6	12	$\sqrt{(6-6)^2+(4-6)^2}$	2.000
9	7,6	14	$\sqrt{(6-7)^2+(4-6)^2}$	2.236
11	8,4	16	$\sqrt{(6-8)^2+(4-4)^2}$	2.000
12	5,5	11	$\sqrt{(6-5)^2+(4-5)^2}$	1.414

(calculation of distance weighted temperature contribution for each used point)

$$\frac{\dfrac{16}{1.414}+\dfrac{12}{2.000}+\dfrac{14}{2.236}+\dfrac{16}{2.000}+\dfrac{11}{1.414}}{\dfrac{1}{1.414}+\dfrac{1}{2.000}+\dfrac{1}{2.236}+\dfrac{1}{2.000}+\dfrac{1}{1.414}}=13.751$$

Thus by distance weighting of the nearest 5 known points, the target site temperature is predicated to be 13.75°C. Using the same calculations it is possible to predict the temperature based on any number of sites being included. Using the 6 sites that seem to surround the target (including site 3 in the above calculation), the prediction is 14.11°C. Using all the available data, the distance weighted interpolation value is 13.34°C.

The data can be interpolated using more sophisticated techniques. The experimental semivariogram of the data can be calculated even though there is relatively little data (Figure 2.9)

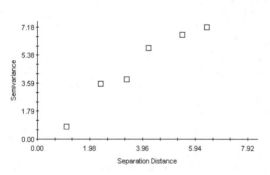

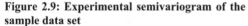

Figure 2.9: Experimental semivariogram of the sample data set

There are a number of models that neatly fit this semivariogram, and given that there is no knowledge about which is best because nothing is known about the geography of the example, we have little information to help us choose the appropriate model. There are a number of statistical descriptors that can assist in choosing the model but many researchers consider these little more than guides, and prefer to use a model that they think intuitively is best. By adopting a "black-box" approach, that is, accepting the recommendation of the software package being used, the best model to adopt is the spherical model, closely followed by a linear-to-sill model (Figure 2.10). In this case, the spherical model will be adopted because it is assumed that temperatures are dependent on those around them and that there is no spatial limit to this effect. If there was a distance beyond which spatial dependence could not exist, then there would be good cause to have a semivariogram model without a sill. In this case the shape of the model make logical sense because there is always some spatial dependence, even at large distances, which the spherical model accommodates even though they may be small.

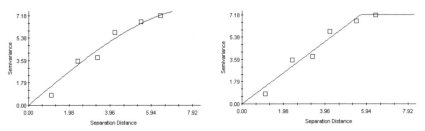

Figure 2.10: (a) the spherical model fit. (b) the linear to sill model fit

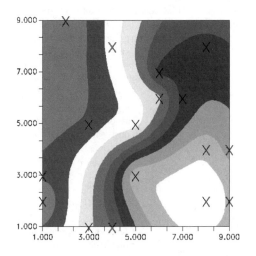

Figure 2.11: Contour map created by point kriging the data using a spherical model

Having calculated the semivariogram model it can be used to krig the data for the purposes of interpolation. Using point kriging (i.e. calculating values for each point over the sample area) a contour map can be constructed from the results (Figure 2.11). With this particular data-set, block kriging (calculating values over a block area, e.g. 2x2, 4x4, 8x8, 16x16) yields very similar results. The kriged interpolation of the temperature data predict a temperature of 14.52^{0}C at the target site with a spherical model. Using a linear model the prediction is 12.88^{0}C. The predicted temperature at the target location 6,4, based on the sample data set varies depending on the interpolation method used. The results are summarised in Table 2.3.

Evaluation of findings
The fact that the data are uniformly distributed means that it should be possible to interpolate to any point in the sample area with reasonable reliability. The clustering of the data towards the centre of the area has resulted in a bias away from a random

distribution, but this is probably an artefact of the data created by the author and not a real feature. It does however serve to illustrate how such data can be interpreted.

Table 2.3: summary of interpolation results

Method	Result (^{0}C)	Value rounded
Nearest value	11 or 16	11 or 16
Nearest 2, distance weighted	13.5	14
Nearest 5, distance weighted	13.75	14
Nearest 6, best surrounding pattern	14.11	14
All data, distance weighted	13.34	13
Spherical model, simple kriging	14.52	15
Linear model, simple kriging	12.88	13

It is quite difficult to evaluate the interpolation results in real terms because the data are artificial, but they do illustrate the methods nicely, and do allow some discussion. Firstly what is noticeable is that all the results are remarkably similar. Once rounded to the nearest whole number, the temperature range predicted is 5^0C. The use of the nearest value available is not good in this case because of the magnitude of the difference between the possible choices. The only justification for using this approach in a real situation is if there are very sparse data, and a mechanistic reason for thinking that the variable changes very little through space. All the remaining methods predict within a 2^0C range. The prediction from the nearest 5 values is higher than that using all the data because the low temperature values at greater distances do exert some influence. It is probably best to restrict the prediction to those samples close by. There may be a case to suggest that site 3 be included in the prediction (i.e. 6 rather than 5 sites) based on the spread of points around the prediction target. When rounded up, the prediction is the same anyway. The kriged estimation based on the spherical model gives the highest predicted value. When rounded up it appears less successful, but in effect the use of the additionally complex method is not necessarily any better in this case. The fact that the linear model of the semivariogram gave a substantially different prediction indicates the sensitivity of this method to the model fitting. Once more it has to be emphasised that kriging should only be used where the data are adequate and a thorough understanding of the system has been developed. A value of 14^0C is probably the best to use for the site 6,4.

References

Bogaert, P., P. Mahau and F. Beckers. (1995): *The Spatial Interpolation of Agro-climatic Data: Cokriging Software and Source Code, User's Manual, Version 1.0b.* Agrometeorology Series Working Paper Number 12, FAO Rome, Italy.

Bonham-Carter, G.F. (1994): Geographic Information Systems for geoscientists: modelling with GIS. Pergamon, Canada.

Bowman, J. 1991. *Acid sensitive surface waters in Ireland.* Environmental Research Unit, Dublin, Ireland.

Clark, I. (1979): *Practical geostatistics.* Applied Science Publications, London.

Cressie, N. A. C. (1985): Fitting variogram models by weighted least squares. *Mathematical Geology* **17**: 563-568.

Cristakos, G. (1992): *Random field Models in the Earth Sciences*. Academic Press, San Diego.

Davis, J. C. (1986): *Statistics and data analysis in geology* (2nd Ed.). J. wiley & Sons, New Yotk.

Ehlers, M., (1997): *Rectification and Registration. Integration of Geographic information systems and remote sensing*. Topics in Remote Sensing 5. Cambridge University Press.

Englund, E. and A. Sparks. (1991): *Geo-EAS 1.2.1 User's Guide*. EPA Report # 600/8-91/008 , EPA-EMSL, Las Vegas, NV.

Farrell, E.P., Cummins, T., Boyle, G.M., Smillie, G.W. and Collins, J.F. (1993): Intensive monitoring of forest ecosystems. *Irish Forestry* 50: 53–69.

Fitzgerald, D. (1984): *Monthly and annual averages of rainfall for Ireland 1951–1980*. Climatological Note No. 7. Dublin, Irish Meteorological Service.

Frogbrook, Z. L. (1999): The effect of sampling intensity on the reliability of predictions and maps of soil properties. In J. V. Stafford (Editor): *Precision Agriculture '99, Part 1*. Sheffield Academic Press, Sheffield, UK. p. 71-80.

Gardiner, M., and Radford J. (1980): *Soil associations of Ireland and their land use potential*. Soil Survey Bulletin No. 36. An Foras Taluntais, Dublin.

Goodale, C.L., Aber, J.D. and Ollinger, S.V. (1998a): Mapping monthly precipitation, temperature and solar radiation for Ireland with polynomial regression and a digital elevation model. *Climate Research* 10: 35–49.

Goodale, C.L., Aber, J.D. and Farrell, E.P. (1998b): Predicting the relative sensitivity of forest production in Ireland to site quality and climate change. *Climate Research* 10: 51–67.

Goovaerts, P. (1997): *Geostatistics for Natural Resources Evaluation*. Oxford University Press, New York.

Goovaerts, P. (1998): Ordinary cokriging revisited. *Mathematical Geology* 30: 21-42.

Goovaerts, P. (1999): Geostatistics in soil science: state-of-the-art and perspectives. *Geoderma* 89: 1-45.

Hettelingh, J-P, Downing, R.J. and de Smet, P.A.M. (1991): *Mapping Critical Loads for Europe*. CCE Technical Report No. 1, RIVM Report No. 259191991. Co-ordination Centre for Effects, National Instiute of Public Health and Environmental Protection, the Netherlands.

Isaaks, E. H. and R. M. Srivastava. (1989): *An Introduction to Applied Geostatistics*. Oxford University Press, New York.

Jordan, C. (1997): Mapping of rainfall chemistry in Ireland 1972–94. *Biology and Environment* 97B, 53–73.

Journal, A. G. and C. J. Huijbregts. (1978): *Mining Geostatistics*. Academic Press, London.

Kitanidis, P. T. (1997): *Introduction to Geostatistics: Applications to Hydrology*. Cambridge University Press, Cambridge, UK.

Legg, C. (1992): *Remote sensing and geographic information systems: Geological mapping, mineral exploration and mining*. Ellis Horwood Limited, New York.

Lillesand, M.T. and R.W. Kiefer. (1994): *Remote Sensing and image interpretation*. Third edition. John Wiley & Sons, Inc.

Longhurst, J.W.S. (1991): *Acid Deposition: Origins, Impacts and Abatement Strategies*. Springer-Verlag, New York.

Matheron, G. (1963): Principles of geostatistics. *Economic Geology* 58: 1246-1266

Matheron, G. (1971): *The Theory of Regionalised Variables and its Applications*. Ecole de Mines, Fontainbleau, France.

McBratney, A. B. and R. Webster. (1986): Choosing functions for the semivariograms of soil properties and fitting them to sampling estimates. *Journal of Soil Science* **37**: 617-639

Nilsson, J. and Grennfelt, P. (Editors). (1988): *Critical loads for sulphur and nitrogen*. Miljörapport 1988: 15. Nordic council of Ministers, Copenhagen.

McGettigan, M.F. and O'Donnell, C. (1995): *Air pollutants in Ireland: emissions, depositions and concentrations*. Environmental Protection Agency, Ardcavan, Wexford.

Molenaar, M. (1990): *Remote sensing as an earth viewing system. Land observation by remote sensing, theory and applications*. Wageningen, The Netherlands.

Norcliffe, G. B. (1977): *Inferential Statistics for Geographers, An Introduction* Hutchinson, London.

Olea, R. A. (1975): *Optimum mapping techniques using regionalised variable theory*. Series on Spatial Analysis. Kansas Geological Survey, Lawrence, Kansas.

Oliver, M. A. (1999): Exploring soil spatial variation geostatistically. In J. V. Stafford (Editor): *Precision Agriculture '99, Part 1*. Sheffield Academic Press, Sheffield, UK. p 3-17.

Ordnance Survey of Ireland. (1993): 1:*100,000 CORINE land cover project (Ireland)*.

Ordnance Survey of Ireland. (1996): *The Irish Grid: A description of the co-ordinate reference system used in Ireland*. Director, Ordnance Survey of Ireland, Government of Ireland.

Rendu, J. -M. (1981): *An introduction to Geostatistical Methods of Mineral Evaluation*. South African Institute of Mining and Metallurgy, Johannesburg.

Ripley, B. D. 1981. *Spatial statistics*. J. wiley & Sons, New York.

Ronald Eastman, J. (1995): *Idrisi for windows user's guide*. Version 1.0, Clark University, MA, USA.

Rosenfeld, A & Kak, A. C. (1982): *Digital Picture Processing (2nd Ed)*. Academic Press, New York. 2 vols.

Schlaepfer, R. (Editor). (1993): *Long-term implications of climate change and air pollution on forest ecosystems*. Vienna, IUFRO; Birmensdorf, WSL.IUFRO Word Series Vol. 4.

Sherwood, M. (Editor).(1992): *Weather, soils and pollution from agriculture*. AGMET, Meteorological Service, Dublin.

Sverdrup, H., de Vries W. and Henriksen A. (1990): *Mapping critical loads*. Miljörapport 1990: 14. Nordic council of ministers, Copenhagen

Wackernagel, H. (1995): *Multivariate Geostatistics: An Introduction with Applications*. Springer-Verlag, Berlin.

Wilkie, D.S. and J.T. Finn. (1996): *Remote sensing imagery for natural resources monitoring: a guide for first-time users*. Columbia University Press.

Chapter 3 Meteorological Data–Types and Sources

T. Keane
Head: Research and Applications (retired)
Met Éireann, Glasnevin, Dublin, Ireland

3.1   Introduction

Agro-meteorology is concerned with the effects of weather on the physical and physiological processes of crops and livestock comprising production agriculture. These processes are influenced by weather, soil, management practices and crop species, but weather is the most temporally variable and least controlled factor, often playing a vital role. The purpose of this chapter is primarily to describe the current Irish meteorological databases, elaborating on their adequacy or limitation for user requirements. In particular, discussion is based on the meteorological data held in Met Éireann and available that from the Meteorological Office in Northern Ireland, with an overview of the station network and the status of the records. The data are reviewed in terms of variables observed, frequency of observation and quality of the information. Reference is also made to factors that determine whether the data from a given station are representative of nearby agricultural areas, their suitability for spatialisation and as input to agro-meteorological models. The information presented in this chapter is typical of that required for any agro-meteorological modelling exercise such as those presented in Chapters 5, 6 and 7.

3.2   Meteorological Database

3.2.1   Agro-meteorological data needs for agricultural production

The key agro-meteorological variables with appropriate specifications needed to meet the goal of crop modelling are:

(a)   Temperature (air and soil; hourly, daily maximum/minimum)
(b)   Precipitation (amount/duration/intensity; form:- rain, drizzle; hail, snow, sleet)
(c)   Radiation (global and net; sunshine duration)
(d)   Moisture content of: air (dewpoint, wet bulb, relative humidity); soil surface (deficit/surplus expressed in mm or hPa); and dew
(e)   Wind (speed/direction at the 10 or 2 m levels)
(f)   Evaporation/evapotranspiration (measured or derived by formula).

To be representative of the broad features of climate, the various variables are measured under recommended conditions, e.g. using standard rain gauges; air temperature taken under shade and at a given height above ground, soil temperature measured at a certain soil depth, and recorded at standard times (Rohan, 1986).

Stations should be well distributed, and sites should have proper exposure with vegetation representative of the surrounding area. If the conditions of representatively are not met, then the data may only reflect the microclimate at the weather station. Other factors of concern to agriculture, e.g. local variations in surface relief, aspect, slope, vegetation, neighbouring water bodies…, also need to be considered (*see* Collins and Cummins, 1996). The appropriate climatic scales and characteristic distances are indicated in Table 3.1 (Guyot, 1998).

Table 3.1: Climatic scales and characteristic distances

Climatic Scale	Characteristic Distances	
	Lowlands/Plains	Uplands/Mountains
Regional (county/district)	100 km	10 km
Topo (farm/field)	10 km	100 m
Micro (field/crop)	100 m	10 m

3.2.2 Reliability and Quality

A key difficulty for modellers is knowing how reliable the meteorological data for the variable are. It is easy to visualize the need for quality data at the location of interest to modellers, but data can often be imperfect or limited in some way and so we are often faced with the decision as to what data are good enough. Even for well maintained standard traditional rain gauges, errors of the order of 3% to 30% can occur due to variation in the coefficient of capture with wind velocity (WMO, 1990). Also some important data used in crop simulation models are not actually measured but derived from other data, e.g. evapotranspiration, vapour pressure, relative humidity, solar radiation (not yet widely measured compared with sunshine), and thus there are increased uncertainties as to reliability. The accuracy of measurement of a variable can often be the limiting factor in the investigation of the quantity to be determined. Hough (1998) estimated that an error of 0.3 ^{0}C in temperature over an eight month period was equivalent to 'about three quarters of a leaf', or an error in the growth stages after crop emergence of 3 to 7 days. Hough also observed that an error of 0.5 hours sunshine daily could lead to ±0.5 t ha^{-1} in winter wheat yields. However, the impact of errors in observations may be diluted somewhat in the smoothing process often associated with interpolation schemes. The recommended minimum accuracy in the measurement of meteorological variables (Villalpando *et al*, 1992) are shown in Table 3.2, but in practice more exact measurements should be aimed for.

Table 3.2: Recommended minimum accuracy of some variables

Variable	Accuracy required in daily values
Temperature (incl. max/min and wet and dry bulb)	$< \pm 0.5\,^0$C
Rainfall	± 1 mm
Solar radiation	10%
(incl. sunshine)	(± 0.1 h)
Evaporation	± 1 mm
Relative humidity	± 5%
Photoperiod	10% (± 0.1 h)
Wind speed	± 0.5 m s^{-1}

3.2.3 The Principal variables

Temperature
Temperature can be a limiting climatic factor in agricultural production in temperate regions. Such is the case when considering the growth of grain maize in Ireland as opposed to silage maize, although new cultivars may make grain maize viable in the future. A comparison of the potential of the different growing seasons in North-West Europe is made from the data in Table 3.3. For crops requiring a sustained air temperature in excess of a threshold of 5.6 ^{0}C, the season ranges from 40 weeks in the south-midlands of Ireland, to 30 weeks in Denmark. At 10 ^{0}C, the range is from 28 weeks in northwestern France to 20 in Denmark. At 15 ^{0}C, Ireland is clearly least favoured, e.g. compare the temperature climate at Kilkenny to that at Cambridge, with 5 and 15 growing weeks, respectively. While weather factors in combination ultimately determine production, temperature is at times particularly important, e.g. at either end of the growing season, and so the length of the frost-free season can be a critical requirement for the production of frost-sensitive crops (Keane, 1986; 1992).

Table 3.3: Approximate length of growing season *vis* a given threshold temperature in Northwest Europe (in weeks)

Threshold temperature	Ireland Kilkenny	England Cambridge	France Le Mans	Netherlands De Bilt	Germany Hannover	Denmark Strömmen
5.6 ^{0}C	40	38	40	35	33	30
10 ^{0}C	24	25	28	25	24	20
15 ^{0}C	5	15	17	13	13	10

Precipitation
Precipitation can be limiting at critical periods for crop growth. Even in Ireland where precipitation is plentiful, the annual rainfall is unlikely to be as important as its distribution during various crop production stages from tilling to reaping. Depending on the phenological phase, a deficiency or excess of moisture may have a significant impact on crops. For example, limited supplies of water may be sufficient from emergence to tillering of wheat (some degree of moisture stress in young plants may be beneficial) but the need for water is greatest during maximum vegetative growth. Also moisture stress at flowering and during grain filling adversely affects growth, giving rise to a reduction in grain yield.

Radiation
Agriculture is a prime example of an industry based on the exploitation of solar energy. For crops with adequate supply of nutrients and water, solar radiation is the only source of energy for physical and biological processes. While radiation is not limiting in many countries, in the context of the climates of North-West Europe, e.g. the Irish climate, it can sometimes be a determining factor for successful production. As a measure of the year-to-year variation in Ireland, the standard deviation (s.d.) in global radiation is about 5 per cent on an annual basis but nearly 10 per cent on a monthly basis. Increased temperature rarely compensates more than a fraction for the yield loss due to the reduced radiation received in dull years. Sunshine hours are often used in crop modelling to represent solar radiation receipt but caution must be exercised when interchanging both in models (*see* 3.3.5).

Relative Humidity
Vapour pressure is one of the more conservative climatic variables despite marked regional discontinuities over short periods of time due to changes of airmass. The formulae for deriving relative humidity and vapour pressure, which take into account the differences in the readings between wet and dry bulb thermometers, are rather complex and beyond the scope of this presentation. Special 'look-up' tables are also available for the purpose. High relative humidity is associated with moist airmasses and upland areas, and night-time values are usually higher than day-time values. For the purposes of agriculture, high values of relative humidity are usually associated with poor drying conditions and the spread of fungal and other pathogen diseases of crops. On the other hand, low values of relative humidity which are detrimental to agriculture, are not normally a factor in temperate climates. The Bourke Irish rules for determining conditions favourable to the spread of potato blight require relative humidity to exceed 90 percent and temperature to be at least 10 ^{0}C for periods of 11 consecutive hours or more (Keane, 1982; also see Chapter 7).

Wind
Ireland is a relatively windy country with frequent gales, mainly in winter. The most notable features of Ireland's wind climate are that the annual mean speed in northwestern coastal areas is more than twice that in the south-midlands (Keane, 1992; Rohan, 1986) and that the frequency of gales is much higher in the northwest. Another characteristic for consideration is that inland at 300 m altitude, the average wind speed is comparable to the exposed coastline at sea level.

Met Éireann archives hourly wind speed and direction, measured at a height of 10 m, for all its (15) synoptic stations. However, most agricultural activity takes place at the 2 m level, where wind speed is less than at 10 m. At a number of climatological stations the run of wind (i.e. daily 24-hour sum) is read at 2 m but this is not archived. While the vertical variation in wind speed depends on surface friction together with airmass stability, the following simplified formula may be used as an approximation to define vertical variation in wind speed, namely:

$$V_u/V_l \cong (h_u/h_l)1/6 \qquad\qquad 3.1$$

where V_u is the wind speed (m s^{-1}) at the higher level (h_u) and V_l is the speed at the lower level (h_l). For the 10 m and 2 m heights, the formula reduces to:

$$V_2 \cong 0.75\,V_{10} \qquad\qquad 3.2$$

Thus, for example, a measured 10 m wind of 4 m s^{-1}, which might suggest undue spray drift, reduces to 3 m s^{-1} at sprayer boom height of 2 m, and thus is considered acceptable for spraying.

Evaporation and Water Balance
As rainfall is highly erratic in intensity and amount (both in time and space), a unit of a month is rarely adequate to properly describe its impact on plant–crop growth, which is mediated through short-term variations in soil moisture conditions. If monthly totals only are available, consideration needs to be given as to whether the precipitation

occurred over a short spell, or if the spell was preceded or followed by a period of (effective) drought. Drought may well result in irrecoverable damage to a vulnerable crop, particularly at critical stages of development. Thus the water available to crops needs to be estimated on a continuous basis over short intervals of time.

A simple water balance may be adequate to provide a good estimate of soil moisture deficit for surfaces covered by short grass. The state of the balance is derived by maintaining running totals of rainfall and evapotranspiration, which are accumulated over intervals of a week to ten days through the growing season or from the start of the year. If the evapotranspiration formula used is sufficiently accurate, then it is preferable to update the water balance on a daily basis. Evaporation (E) from a water surface and potential evapotranspiration (PET) from a grass crop are generally regarded as standard reference quantities similar to temperature or radiation and thus are normally observed or calculated. PET is sometimes measured by special grass (or other crop) canopy-covered lysimeter. However, because of the lack of observing stations, PET is now normally derived by formula, e.g. by the so-called Penman combination equation (i.e. computation of the separate radiation and wind contributions to evaporation) (Connaughton, 1967; HMSO, 1967). Such a formula requires as input the various weather variables observed at the synoptic and many climatological stations.

An evapotranspiration equation with a physical basis, known as the Penman-Monteith equation, is widely used and recommended by the Food and Agriculture Organization (FAO) of the United Nations (Smith *et al.*, 1991). It also forms the basis for the MORECS (Meteorological Office Rainfall and Evaporation Calculation System) method in the UK (Hough *et al*, 1996; Hough and Jones, 1997). The original Penman-Monteith equation takes the following form:

$$\lambda\, ET \;=\; \frac{\delta\,(R_n - G) + \rho\, c_p\,(e_a - e_d)\,{}^{1}\!/_{r_a}}{\delta + \gamma\,(1 + r_c/r_a)} \qquad\qquad 3.3$$

where ET is the rate of water loss through evapotranspiration (kg m^{-2} d^{-1}) and λ, latent heat of vaporisation (or λET, latent heat flux of evaporation (MJ m^{-2} d^{-1})); R_n, Net radiation flux at surface; G, soil heat flux; δ, slope of vapour pressure curve; ρ, atmospheric density; c_p, specific heat of moist air; $(e_a\text{-}e_d)$, vapour pressure deficit; r_c, crop canopy resistance; r_a, air dynamic resistance; γ, phychrometric constant (see Table 3.4 for units).

The Hough *et al* (1996) description of MORECS provides considerable detail on the variety and range of variable values in use in the system. In their absence, e.g. canopy resistance for different crops, the FAO suggests the useful approximations presented in Table 3.4.

The diurnal amounts of incoming extra terrestrial radiation at various latitudes and seasons, may be obtained by formula. The formula for calculating the extraterrestrial radiation on a horizontal surface (Iqbal, 1983) has the following form:

Table 3.4: Approximate variables recommended by FAO for use in Penman-Monteith equation

Variable	Approximation	notes [Units]
λ	$\cong 2.501-(2.361x10^{-3})T$	[T, ^{0}C][MJ kg^{-1}]
δ	$\cong 4098e_a/(T+237.3)^2$	[kPa ^{0}C^{-1}]
γ	$\cong 0.0016286\ P/\lambda \cong (0.66)P$	(kPa), atmospheric pressure [kPa ^{0}C^{-1}]
ρ	$\cong 3.486\ P/T_{kv}$, where $T_{kv} \cong (T + 275)$	virtual temperature [kg m^{-3}]
c_p	$\cong 1.013$	[kJ kg^{-1} ^{0}C^{-1}]
e_a	$\cong 0.6108\ exp(17.27T)/(T+237.3)$	[kPa]
e_d	$\cong e_a(T_{mean})(RH_{mean}/100)$ $\cong 0.6108\ exp^{((17.27Tmin)/(Tmin+237.3))}$	(daily mean if hourly values unavailable) [kPa] (if RH absent, assumes saturation at minimum temperature)
r_c	$\cong 200/LAI$	where for grass suggested Leaf Area Index, $LAI \cong 2.88$ [s m^{-1}]
r_a	$\cong 208/U_2$	where U_2 is the 2 m wind (see equation 3.2) [s m^{-1}]
R_n	$\cong R_{ns}\downarrow + R_b$	$(R_{ns}\downarrow$ is net incoming short-wave radiation, $R_b\uparrow$ net outgoing long wave radiation) [MJ m^{-2} d^{-1}]
R_{ns}	$\cong 0.77((0.18+0.55n/N)R_a$	(where R_a, extra terrestrial radiation; n, sunshine duration; N, length of day in hours; $0.77 = 1-\alpha$ where α (0.23) is canopy reflection)
$R_b\uparrow$	$\cong -(0.9n/N + 0.1).(0.34 -0.14\sqrt{e_d}).\sigma.\ \frac{1}{2}(T_{kx}^4 + T_{kn}^4)$	T_{kx} and T_{kn} are the maximum and minimum day temperatures, [^{0}K]
σ	$\cong 4.903.1$	Stefan Boltzmann constant [MJ m^{-2} K^{-4} d^{-1}]
G	$\cong 0.38\ (T_{day\ n} - T_{day\ n-1})$ $\cong 0.07\ (T_{month\ n} - T_{month\ n-1})$	for mean daily temperature fluctuations or for monthly temperature fluctuations (Since the magnitude of G (soil heat flux) over 10-30 days is relatively small, it normally can be neglected and thus G $\cong 0$) [MJm^{-2} K^{-4} d^{-1}]

*Source: Smith *et al*, 1991 (Note: 3.6 MJ = 1 kW hr)

$$H_o = (24/\pi)\ I_{sc}E_o\ cos\phi\ cos\delta'\ [sin\omega_s-(\pi/180)\omega_s\ cos\omega_s] \qquad 3.4$$

Where H_o is the extraterrestrial daily irradiation incident on a horizontal surface (MJ m^{-2} d^{-1}), I_{sc} is the solar constant (1367 Wm^{-2}), ϕ is the latitude in degrees, δ' is the solar declination (degrees), E_o is the eccentricity correction factor (dimensionless) given by the approximate formula (Iqbal, equation 1.2.3):

$$E_o = 1+ 0.033\ cos[(2\pi d_n/365] \qquad 3.4a$$

where d_n day number, and ω_s is the sunrise hour angle (degrees) given by:

$$\omega_s = cos^{-1}(-tan\phi \; tan\delta')$$ 3.4b

The solar declination, δ', is the angle (degrees, north positive) between lines joining the centres of the sun and the earth to the equatorial plane, which changes daily. Although calculated by a complex formula, δ' may with sufficient accuracy be simplified to the following expression (Perrin de Brichambaut, 1975; cit. Iqbal, 1983):

$$\delta' = 23.45 \; sin[(360/365)(d_n + 284)]$$ 3.4c

Alternatively, incoming extra terrestrial radiation may be interpolated from Table 3.5 (adapted from Budyko, 1974). The lack of a simple relationship between sunshine hours and solar radiation (see 3.3.5) means the Penman-Monteith equation is suitable for determining 10-day or weekly totals but can also be used for any duration, i.e. daily values, provided the appropriate input data can be provided.

Table 3.5: Solar radiation–R_a incident on upper atmosphere (MJ $m^{-2}d^{-1}$) on specified dates

Latitude (degrees)	Feb. 22	Mar. 4	May 21	June 6	Aug. 22	Sept. 8	Nov. 23	Dec. 8
60 N	6.28	19.26	34.75	42.29	34.75	19.26	6.28	2.09
50	12.56	24.70	37.26	42.71	37.26	24.70	12.56	7.37
40	18.84	29.73	39.36	42.71	38.94	29.31	18.42	13.82
30	24.70	33.50	40.20	41.87	39.78	33.08	24.28	21.00
20	29.73	36.43	39.78	40.20	39.36	36.01	29.73	25.96
10	34.33	38.10	38.52	37.68	38.10	37.68	33.91	31.82
0	36.43	38.52	36.01	33.91	36.01	38.10	36.43	37.68
10	40.20	38.10	32.66	29.73	32.66	36.43	40.20	40.20
20	41.87	36.43	28.47	24.28	28.05	36.01	41.45	43.13
30	41.87	33.50	23.45	16.75	23.45	33.08	41.45	44.80
40	41.04	29.73	18.00	12.98	17.59	29.31	40.61	45.64
50	39.36	24.70	11.72	7.12	11.72	24.70	38.93	45.64
60 S	36.42	19.26	5.86	2.09	5.86	19.26	36.43	45.22

adapted from Budyko (1974)

An alternative simplified formula, known as the Priestly and Taylor formula, which describes evapotranspiration loss of wet land surface surprisingly well, has the following form:

$$\lambda ET = \alpha[\delta/(\delta + \gamma)](R_n - G)$$ 3.5

where as above λET, latent heat flux density; R_n, net radiation; G, soil heat flux, δ, slope of the saturation vapour pressure-temperature curve at the wet bulb temperature; γ, psychometric constant; and α, an adjustment factor.

While the value of α (~ 1.2 - 1.3) varies with type and maturity of vegetation, time of day, and weather conditions, a value of 1.26 is normally accepted for potential growing conditions (Pereira and Villa Nova, 1992). In temperate latitudes, G is considered to be small for grassland and thus neglected.

As net radiation is not widely measured, and difficult to calculate (see Penman-Monteith formula given above), a further simplification for well watered grassland is as follows:

$$\lambda ET = C[\delta/(\delta + \gamma)]K\downarrow \qquad \qquad 3.6$$

where $K\downarrow$ is the incoming short wave (global) radiation (MJ m^{-2} d^{-1}). A factor C of 0.65 fits reasonably well potential-evapotranspiration (PET) from grass according to Hooghart (1987). Global radiation may be obtained by direct measurement or derived by using the Ångstrom formula (see, for example, equation 3.7).

3.2.4 Data Needs

For most crops, growth occurs over a period of about 100-120 days. Using simple crop models or biophysical indices, the minimum climatological data set must at least be sufficient to provide an adequate explanation of the initial and subsequent states of growth. The appropriate time-averaged interval will vary with crop, and with model and location. Normally the time unit for analyzing climatic data is chosen in such a way that the calculation can reasonably reflect the more important phenological stages and the various influences of weather, without undue computational load (Table 3.6). A period such as a week or 10 days as an averaging unit is often considered appropriate although some mechanism must be used to take into account the effects of significant shorter period events, e.g. overnight severe frost, stress effects from sudden disease epidemics on crops.

Table 3.6: Time intervals normally used in models

Model	Time Interval
Production/Yield	10-day summary of daily values
Crop diseases	Mostly daily values some hourly and weekly summaries
Animal Diseases	Monthly/seasonal climatologies

3.3 Network and Status of Meteorological Records

The physical factors determining the sphere of representivity of a station are topography, altitude, proximity to the sea and the natural barriers presented by high ranges and mountains. An important factor also is the zone of transition from maritime to inland climates. While climatological gradients from sea to inland can often be smooth, strong variable gradients do occur over a range of 20 km from the sea, with sharpest changes confined to the first 6-10 km (*see* section 3.4(b)). The extent of the coastal zone, however, depends on the prevailing wind (e.g. south-westerlies in Ireland), the synoptic situation (whether associated with stable high pressure or broad warm sector conditions) and the variables considered.

Discussion in the following section considers the various types of observing stations in relation to areal coverage, representivity and period of record. Such information is necessary, for example, to determine adequacy for regional climatologies, validation of crop models and to simulate plant development and disease progress. Guyot (1998) suggested the following separation between observing stations (Table 3.7) as desirable. Such spacing is unlikely to be achieved in the short-term even in the more advanced countries, until programmes for increased use of low cost automatic stations are more widely put in place.

Table 3.7: Recommended Separation between Observing Stations

Element	Station Separation (km)
Rainfall	15
Air Temperature	30
Other regular (synoptic) variables	60
Sunshine Recorders	70
Solarimeters (Radiation)	150

3.3.1 Network of Rainfall Stations - The Irish example

Distribution: There are some 600 'rainfall stations' in the Republic of Ireland and a further 200 in Northern Ireland. Rainfall amounts are measured once daily except for a small number of stations which have rain-recorders providing continuous measurement throughout the day. Elevations of the rainfall stations vary from sea level to over 800 m with most below 300 m. Where no local records of rainfall exist, the choice of the most appropriate representative rainfall station needs careful consideration as the difference in rainfall accumulations between stations can sometimes be considerable even over short distances. The monthly and annual rainfall averages for rainfall stations are normally published by each national meteorological service. For example, the climatological statistics for Ireland are entitled: *Monthly and Annual averages of Rainfall for Ireland 1961 - 1990* (Fitzgerald and Forrestal, 1996). In due course these data are likely to be available on the Met Éireann web site to authorised users. Corresponding data for Northern Ireland are available from the Meteorological Office, Belfast.

Figure 3.1 shows the distribution of rainfall stations in Ireland. The network is in a constant state of flux in that some stations are closing and new ones opening over any given period. However, there is a large body of permanent stations. The rainfall network coverage is the most representative network *vis-a-vis* altitude compared with any of the other variables which are measured.

3.3.2 Synoptic Station Network

Traditionally the Irish synoptic station network (Figure 3.1) measure hourly values of the standard meteorological elements, e.g. atmospheric pressure, air dry and wet bulb temperature, wind, precipitation, daily maximum and minimum temperature, sunshine and/or global radiation and soil temperatures. Manual synoptic stations also record present and past weather conditions, cloud amount and type. Derived variables such as relative humidity (RH) are calculated on an hourly basis. For the years 1992-98, breaks occurred in the continuity of recording of some variables at many inland synoptic stations but not usually those of importance to agriculture. The network has recently been brought back to full routine operation by the installation of automatic 'synoptic' stations operating in parallel with manual observations. In Northern Ireland, apart from Aldergrove, many 24-hour stations have been of the automated type for a number of years.

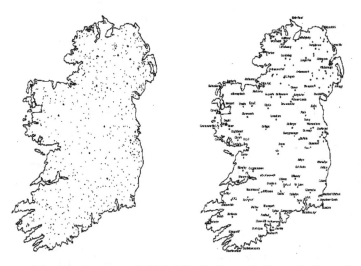

Figure 3.1: Distribution of Rainfall Stations (left) and Synoptic and Climatological Stations in Ireland

Automatic recording weather stations will be more widely used in future years. With their use, however, there is a loss of some traditional elements, e.g. the instrument for the measurement of sunshine by automatic weather stations is not quite comparable to the traditional sunshine recorder. On the other-hand, the recording of new and useful elements such as global radiation and duration of (leaf) wetness is possible. Despite the various software checks put in place, automatic stations still need vigilant monitoring for the early detection of malfunctions and errors.

Distribution: About half of the synoptic stations in Ireland are situated on or near the coast, which for the purposes of agriculture may not be fully representative of areas beyond a 20-30 km coastal strip. Also, some synoptic stations are located at airports, and thus do not fully reflect conditions in the surrounding district. Shannon Airport is a station in point, where the annual degree-day totals suggest that the thermal conditions there are some 10% warmer than the surrounding region. Most inland synoptic stations are at low altitudes, Cork and Knock Airports being the most elevated at 154 m and 203 m respectively.

Though often in the vicinity of suburban housing, many synoptic stations are nevertheless acceptably representative of the major surrounding regions as roughly indicated in Figure 3.2. Coastal stations are representative only for narrow strips whereas inland stations are normally representative of much more extensive areas.

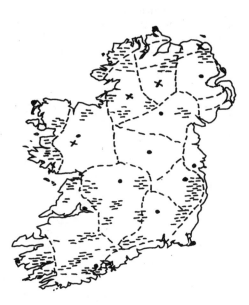

Figure 3.2: Approximate areas of Ireland (delineated by dash lines) whose climate is broadly represented by the enclosed synoptic stations (indicated by a dot) or automatic station (X); high ground is shown as hatched areas

Upland areas for the most part are not well catered for by lowland synoptic stations, and these areas are depicted by stippling. The minimum set of synoptic stations which broadly represent the major Irish agricultural regions is indicated in Table 3.8.

A summary of the monthly mean and extreme climatic data for each of the stations may be obtained from meteorological publications (e.g. Keane, 1985). Where daily and hourly data are required for a wide network of stations, recourse may be made as appropriate to the data archives in Met Éireann, Dublin or the Meteorological Office, Belfast. For those requiring data in other countries, contact should be made to the relevant National Meteorological Service.

Table 3.8: Synoptic Stations and the Major (inland) Agricultural Regions

Synoptic Station	Region
Claremorris	West/Northwest
Cork Airport	South/Southwest
Kilkenny	South-Midlands/Southeast
Mullingar	Midlands
Clones	North-Midlands and South Ulster (N.I.)
Aldergrove	Northern Ireland

3.3.3 Climatological stations

Manually operated climatological stations (also included in Figure 3.1) provide daily readings of maximum and minimum air temperature, 0900 Coordinated Universal Time (UTC) air temperature, dew point and daily rainfall, and at many stations, sunshine hours, grass minimum and soil temperature (at least at one or two depths, e.g. 100, 300 mm). It is to be noted that synoptic stations also qualify as climatological stations. Some climatological stations, and all (manual) synoptic stations, provide 24-hr rainfall recorder traces. Even in most temperate regions, there can be a dearth of upland stations although, in recent times, this situation is being remedied somewhat by the employment of automatic weather recorders in these areas. Automated recorders provide data at any predetermined interval. The ready availability of data from climatological stations, has traditionally been held up by the procedure for once monthly return of completed forms for quality control before archiving of the data. With expanded use of the automatic weather stations, and greater use being made of e-mail facilities, such data are likely to become available to users in near real-time in future years. In the Irish station network, there are about 84 climatological stations, of which only a limited number (40) are published monthly in Met Éireann's 'Monthly Weather Bulletin'; all are archived however. Over 50 climatological stations are operated and archived in the Meteorological Office in Northern Ireland.

3.3.4 Evaporimeter Instruments

Evaporimeter instruments, such as the Class A Pan or grass covered lysimeters, for measuring evaporation and evapotranspiration respectively, are less used operationally in Europe in recent years. In Ireland, however, a distributed number of some 20 weather stations continue to measure daily evaporation using the so called Class A open water pan evaporimeters. A small number of stations use Thornthwaite grass covered lysimeters to measure evapotranspiration from a short grass surface. These can be useful as 'ground truth' instruments, particularly for field trials.

There are, however, many potential causes of error with use of the Class A Pan, such as water loss through splashing in high wind or sometimes by bathing birds (a fine netting may be used to protect from the latter). While daily values of evaporation from the Class A Pan have to be taken with caution, they provide useful estimates for longer periods such as a month. Thus, because of the uncertainties in the daily readings, only monthly Class A Pan evaporation values are archived in the Met Éireann database. The need for close monitoring, and the difficulty of automation means the number of such evaporimeters is likely to continue to decrease as more weather stations become automated.

Conversion Factor for Large-Scale Evaporation
To account for the additional solar heating effect on a small, raised volume of water compared with a large body of lake water, open pan evaporation may be converted to shallow lake evaporation by applying a recommended but very approximate reduction factor of 0.75 to 0.80. Furthermore, by applying an additional factor of 0.75 (to account of crop resistance to transpiration), pan evaporation may be converted to approximate monthly rates of potential evapotranspiration (PET) for a short grass surface, i.e. giving a combined coefficient of 0.75 (or 0.80) x 0.75 ≅ 0.5/0.6. However, as Class-A pans are affected more by the advective component of evapotranspiration than by the radiative component, and as the proportional contribution to PET differ for crops and Class-A, the coefficient may in fact vary within a range approximating 0.50 to 0.75, the lower values being associated with strong winds (Guyot, 1998; Doorenbos and Pruitt, 1977).

3.3.5 Radiation and Sunshine

Solar radiation
Solar radiation is not measured as widely as sunshine hours. For example, only half of the synoptic stations in Ireland directly record radiation at present. With increased use of automatic weather stations (AWS's) this ratio is likely to improve greatly. Global radiation is the most widely measured radiation variable although net radiation is perhaps more relevant to crop growth. Traditionally, global radiation has been calculated from sunshine hours (e.g. equation 3.7) and net radiation is calculated from global radiation by applying a correction for surface temperature to estimate the amount of long-wave radiation lost to the sky (e.g. see $R_b\uparrow$ in the Penman-Monteith formula in section 3.2.3, Table 3.4).

A quality controlled archive of Irish radiation data (dating from the 1960's) containing hourly and daily values of global (7 stations), diffuse (5 stations) and net radiation (measured at Valentia Observatory and Kilkenny) is held in Met Éireann. The data are entered into the archive within three to four months of observation, but in future years the data should be available more promptly.

Sunshine
Daily and hourly values of sunshine are available on the Met Éireann database for a network of 50 synoptic and climatological stations. Provisional first line quality control is available for 'synoptic' stations on the day after recording. Full quality controlled sunshine data from all climatological stations become available during the month following observation.

Estimation of Global Radiation
The formula commonly used to estimate global radiation at stations where sunshine hours are measured is known as the Ångstrom formula. While different studies give various values of the parameters used, a useful form of the equation for temperate climates is given by (HMSO, 1967):

$$R_{gb} = R_a(0.18 + 0.55n/N) \qquad\qquad 3.7$$

where R_{gb} is the global radiation (in Joules or watt-hours per unit area); R_a is the radiation received at the outer atmosphere; n is the daily sunshine (hours); and N is the maximum possible daily sunshine for time of year. R_a varies regularly through the season and may be obtained from appropriate ephemeral tables, such as given in Table 3.5, or by formula (*see* section 3.2.3). Equation 3.7 is valid for monthly intervals but may usefully be used for shorter periods such as 10 days or a week.

An estimation of the global radiation at a given location (weather station) may be made from knowledge of local sunshine hours, and global radiation and sunshine hours at a nearby station. The following formula (albeit with parameters somewhat different to equation 3.7) has been used to calculate such values:

$$R_{gb, st1} = R_{gb, st2} (0.21 + 0.60 (n/N)_{st1}) / (0.21 + 0.60 (n/N)_{st2}) \qquad\qquad 3.8$$

where, for example, sunshine only is available at weather station st_1, and both sunshine and radiation are measured at station st_2. From a study of five years of data, in the growing season, March to September, and using Birr as st_1 and Kilkenny as st_2, a correlation of 0.98 with standard error of estimate (SEE) of 3.5% was derived for 30-day averages whereas as a correlation of 0.96 and SEE of 7.1% was obtained for 5-day averages (Keane, 1998).

A Note of Caution
It should be noted that crop growth simulation models validated using sunshine hours may not necessarily produce sufficiently comparable outputs if directly measured solar radiation is used as input data instead of sunshine. This results from the assumption made in deriving the model regressions that a simple linear relationship exists between solar radiation and sunshine hours, while in reality this is not the case. The problem is exemplified in a correlation analysis between sunshine and radiation for a station, by the considerable spread of the data points about the regression line. For a given daily sunshine total, the corresponding radiation depends on the atmospheric turbidity and the time of day in which the sunshine occurred, e.g. a comparison made between noon and early morning will give vastly different radiation amounts even if the same sunshine amount is recorded.

3.3.7 Other Variables

Atmospheric pollutants
Sampling of monthly rainfall for chemical analysis has been carried at a number (8) of synoptic stations in Ireland since 1958. The pH measurement is the simplest and most commonly used index of acid rain; electrical conductivity provides a measure of the total ionic concentration of the soluble ions present; other analysis carried out on a monthly and daily basis include SO_4-S, Sulphate, NO_3-N, Nitrate, Cl, Na, Ca, K, Mg and NH_4-N,Ammonium. The data become available within a month or two after observation. A long-term archive has now been established so that it is possible to obtain past data for varying periods.

Ozone measurement
Ultra violet radiation measurements are made at Valentia Observatory, Malin Head and at Mace Head. The Valentia and Malin Head data are stored in Met Éireann. Ozone is monitored at Valentia Observatory using a spectrometer to measure total column amounts to obtain vertical ozone profiles. Special ozonesonde ascents are also made in winter. The National Radiological Board measures ozone at surface level at a number of locations. Also surface ozone is monitored at Oakpark, Carlow and at some other sites in Cork and Tipperary during the growing season.

3.4 Interpolation of the Data

Meteorological data used for crop growth simulation models or for pest/diseases risk assessments are either directly observed or interpolated from a set of neighbouring values. While crop models for a particular farm are best run on data from an on-farm weather system, data from the nearest meteorological station may also be used if the station is similarly sited. However, it is often the case that at a given point an appropriate interpolation scheme needs to be used. Interpolation methods enable variables to be estimated at grid points with a high spatial and temporal resolution. An example of a complex interpolation scheme is the kriging method. The method employs a geostatistical model of suitable complexity that describes spatial patterns. The kriging method is described in most textbooks on interpolation and GIS (Cressie, 1993; Kitandis, 1997; also see Chapter 2).

Indeed in many GIS systems, meteorological, soils and topographical data are interpolated to grid points or blocks using an appropriate scheme. Such schemes need to incorporate meteorological data received from many sources and with varying characteristics. As well as conventional (manual) observations, data are received from automatic and remote sensing sources with different time-scales. In addition, there is now a possibility to use grid point data from numerical prediction analysis schemes (see Chapter 8) in such interpolation. The various sources of data need to be harmonized within whatever scheme that is employed.

At times, however, a simple method for interpolating to a particular location, monthly (or perhaps weekly) data from a limited number of surrounding stations may be adequate. One such method entails first determining differences from normal of the variable over a stated period at each station, interpolating these deviations to the point of interest, and where the normal for the new location or grid point is known, then using simple weighted averages, converting the interpolated departures to actual values of the variable. This procedure might use a simple weighting function such a one which depends on the inverse distance of the observation point to the place of interest.

More generally, the weighted difference interpolation scheme uses the recorded variable at a number of nearest-neighbour stations, preferably in different directions and with similar exposure, elevation and aspect (i.e. on same side of a ridge or mountain range). To interpolate the data to a given point, a weighted mean of the surrounding station values may be obtained, for example, by the following formula (Meteo Consult BV, 1991): (note: $W_t = 100 - D/2$, where distance D is in km, (i.e. positive up to distances of 200 km) and W_t greater or equal to 0):

$$Y_{est} = (Y_{st1} \times Wt_1 + Y_{st2} \times Wt_2 + Y_{st3} \times Wt_3)/(Wt_1 + Wt_2 + Wt_3) \qquad 3.9$$

W_t is the weight to be applied to each of, say, three station used in the calculation, Y_{est} is the estimated value of the variable at the point of interest, Y_{st1} is the variable at station 1 with weight Wt_1, the latter being determined by its distance from the point to be calculated. The worked example in Chapter 2 illustrates this method.

The distance-weighted method has many advantages over other more complex methods. It is generally preferred because of comparative accuracy, simple implementation and ease of adaptation for missing values of unequal accuracy (Rijks et al, 1998).

Rainfall
Due to great variability, rainfall should be calculated by an appropriate interpolation method. However, where recourse to such a scheme is not possible, it can sometimes be adequate to assume that rainfall at the location of interest is the same as that at the

nearest measuring station, if not too distant, or preferably by the method described above. Except during periods of convective precipitation resulting in localized showers, the one-station method could on occasion be sufficient if the stations are is not more than, say, 20-30 km distant from one another, the accumulating period is not too short and where high hills or mountain ranges do not intervene.

The great variability in rainfall with altitude means a two-step interpolation method for increasing elevation is necessary rather than using the nearest (neighbouring) station alone. If there is a large difference in topographical elevation, one can take as a working assumption that precipitation increases by 10-15% per 100 m increase in altitude, the larger rate occurring on rising ground facing the prevailing wind and least variation on the leeward slopes experiencing a 'rain shadow' effect. At 300 m, the rainfall is usually 40% higher than the adjacent lowlands. This rate of increase is valid only for rainfall totals over extended periods, e.g. monthly and annual amounts, when the random effects of airmass differences in precipitation type between localities are smoothed out. The rule may sometimes be usefully employed with caution for shorter periods of 10 days.

Smith (1976) provides a useful but generalized formula for an increase in annual precipitation with altitude as:

$$\Delta RR = (0.315 \; RR_{an.} - 119) \; mm \qquad 3.10$$

where $RR_{an.}$ is the mean annual precipitation and ΔRR is the increase in annual amount per 100m. This formula essentially gives the altitude effect ranging from 17% in areas of low (800 mm) annual rainfall to 23% in areas of high (1400 mm) annual rainfall per 100 m.

Temperature
(a) Interpolation between stations
The interpolation of mean air temperature between observing stations can normally be carried out in similar fashion to rainfall, with best results obtained from the more sophisticated schemes. Other difficulties that need consideration in the case of temperature, e.g. (a) the period of integration may have more than one prevailing weather type such as a cold spell followed by a mild spell in the same period; (b) maximum temperature also depends on the local solar radiation (sunshine) received; (c) minimum temperature can generally only be inferred as the actual value depends to some degree on the local topographical air drainage features; and (d) grass minimum temperature has a very localized representivity.

(b) The coastal effect
In a moderate wind and uniform airmass, the change in thermal conditions from coast to inland is gradual, in which case a linear interpolation of temperature between coastal and inland stations is sufficient. This is particularly true on most occasions for western coastal areas of Ireland. During the 'summer half' of the year, interpolation in eastern coastal areas of monthly mean air temperature (°C) near some coasts is reflected by the approximate form:

$$\Delta T_{mean} \cong 0.05 + 0.035D \qquad 3.11$$

for distances (D) ≤ 20 km from the coast (D = 0 at coast).

The mean daily maximum, T_{mx} (°C) air temperature increases inwards during the growing season from the coast, and mean daily minimum, T_{mn} (°C) air temperature decreases according to power law of the type:

$$\Delta T_{mx} \cong 0.5 \ D^{0.6} \quad (D \le 20 \ km, \ D \ne 0)$$ 3.12

and

$$\Delta T_{mn} \cong - \ 0.25 \ D^{0.4} \quad (D \le 20 \ km, \ D \ne 0)$$ 3.13

In the latter case the higher temperature occurs near the coast. For points further inland, the temperature may be interpreted in the normal way between the modified value at 20 km and nearest inland weather stations.

(c) Adjustment of temperature for altitude
The lapse rate in the mean air temperature, ΔT, is often taken as 0.6 °C decrease in temperature per 100 m increase in altitude. The actual lapse rate, however, depends on the stability of the air in a given synoptic situation, the larger rates occurring in unstable airmasses. While the air temperature, including day time maximum, is lower on the higher ground compared to the lowlands, in contrast the higher minima occur on the elevated ground, e.g on clear calm nights a negative lapse rate develops as cold air drains into the valleys. A complex situation can occur in the presence of shelterbelts or if the ground at the base of the valley is slightly raised due to the presence of a local mound or small hill. On a skyward radiating night (e.g. clear, calm night), locally elevated features will experience less intense frost compared with their surrounds.

(d) Soil Temperatures
For seed germination, soil temperature is more important than air temperature. During sunny spells, soil temperature at seed depth can be much greater than the air temperature. Thus the use of air temperature in models can underestimate seedling emergence rates. The daily minimum soil temperature (300 mm depth) may be interpolated from the 0900 hours air temperature by the following equation (Hough et al, 1998):

Minimum (mn) soil temperature (°C):

$$T_{mn \ (soil)} = T_{air \ (09 \ hrs)} - 1.5$$ 3.14

Hough also indicated the following type formula for deriving maximum soil temperature having the form:

$$T_{mx \ (soil)} = A + B(T_{mx \ (air)} - T_{mn \ (air)}) + C \ x \ (daily \ radiation)$$ 3.15

where the coefficients A, B, C need to be regionally determined by regression. The mean daily soil temperature will thus be determined as:

$$T_{mean \ (soil)} = 1/2 \ (T_{mn \ (soil)} + T_{mx \ (soil)})$$ 3.16

Sunshine

A simple interpolation of sunshine hours from surrounding stations to a location can often provide sufficient accuracy. However, the interpolation procedure should take topographical features into account so that extreme values are not overlooked, e.g. an increase in cloudiness on the windward side of mountains or decrease on the leeside. During the summer period, more sunshine hours are normally received at coastal stations compared to sites a short distance inland. Thus, when interpolating in coastal regions, greater weight should be given to inland stations, applying gentle gradients in areas well away from the coast and confining the sharper gradients to coastal strips up to about 10 km. For example, if the daily total sunshine hours were 10 hours at the coast and 6 hours at 50 km inland, a steep gradient of 0.2 hour km^{-1} seems reasonable over the first 10 km whereas 0.05 hour km^{-1} would reflect better the remaining gradient. Sometimes the inclusion of 'bogus' sunshine observations are considered as a means to force the calculations in such areas or in data void regions (Hamilton *et al.*, 1987).

3.5 Estimation of other Variables

Relative Humidity

Owing to the non-linearity of the relationship between saturation vapour pressure and temperature, different results can be obtained depending on the interpolation procedure adopted. For example, relative humidity (RH) calculated at a point and then interpolated to a second (or grid) point may yield quite different results to that obtained by first interpolating the wet- and dry-bulb temperatures to the second point before the calculation. It not clear as to which is the better procedure for interpolation (p. 44, Hough *et al*, 1998). For a given elevation, mean RH is, in general, spatially conservative. Therefore for the purposes of extrapolation, and as an approximation over averaging periods of ten days or more, RH can be taken to vary smoothly between inland stations. With elevation, RH increases in unsaturated air by about 3% for each 100 m.

Evapotranspiration

Similar to relative humidity and vapour pressure, there is no clear guidance as to whether potential evapotranspiration (PET) should first be calculated at the weather station before interpolation to another (grid) point, or if the station weather variables should be first interpolated to the point before the calculation of PET. However, no systematic difference is apparent between the methods.

The surroundings of a weather station can influence estimates of PET. Major modifying effects are adjacent large areas of water, or the presence of vigorously transpiring crops or trees. Crop canopy roughness reduces wind speed and increases the absolute humidity of the air, especially in summer, with consequential reduction in the estimated PET. Airport sites on the other hand show increased PET due to reduced vapour pressure and also higher windspeeds at these locations. Also higher values of PET occur near coasts because of the increased sunshine and wind on the coasts. The spatial variation in the computed (Penman) PET values for inland areas are generally smaller than for most meteorological variables.

The biggest areal change in evapotranspiration occurs with altitude; a reduction of the order of about 3 mm per 100 m increase in altitude on the monthly PET during the growing season, March to September, is perhaps a conservative estimate. Taken in combination, the lower temperature, lesser sunshine (greater cloudiness), increased relative humidity and higher rainfall with altitude contribute to lower PET values and more persistent wetness in upland areas (Keane, 1986).

Soil Moisture

As a general indicator, average daily PET varies from 0.5 mm in December and January to 3 mm or more in the period, May to July. Apart from the prevailing weather conditions and time of year, the actual value depends on the exposure and location of the site, whether inland or coastal. Rainfall normally decreases in summer and PET progressively increases so that a soil moisture deficit gradually develops. However even in summer months, after prolonged wet spells resulting in high soil moisture levels, the state of moisture can easily return to field capacity or sometimes to saturation levels. Such conditions often result in poor land trafficability, a raised water-table with unseasonable difficulties for (forage) harvesting (Brereton and Korte, 1997).

Three cases may be considered:

(i) Soil Moisture Surplus:- When soil moisture is above field capacity (FC; state at which the soil is holding maximum amount of water after a period of drainage), the excess water in the soil is depleted at a daily rate by the sum of PET and the percolation/drainage rate (Brereton,1989). For practical purposes, the percolation rate in medium-heavy soils may be taken simply as 3 mm per day. In the calculations of the water balance, as an approximation in the absence specific information on run-off characteristics and soil type, it may be useful to assume that soils exceed field capacity to a maximum of 10 mm, and above that state, the excess rainfall runs off over-ground. Such an approach should be carefully evaluated prior to application.

(ii) Moderate Soil Moisture Deficits:- When the soil moisture falls below FC, a soil moisture deficit (SMD) exists, and further moisture depletion from the soil/crop takes place for some time at the full PET rate;

(iii) Large Soil Moisture Deficits:- When SMD is greater than 40 mm (the precise value depending on crop), soil moisture depletion takes place at a restricted rate known as Actual Evapotranspiration (AET). To calculate AET, the Aslyng scale (Aslyng, 1965) may be used for a grass crop as follows:

$$AET = PET\ (120 - SMD^*)/(120 - 40)\ \ mm \qquad\qquad 3.17$$

where SMD > 40 mm, and SMD* is the accumulated soil moisture deficit at the beginning of each period (daily, weekly or 10-day). This leads to an updated SMD as follows: SMD = SMD* + (AET - RR) where RR is the total rainfall (mm) for the period. The continued reduction in soil moisture (increase in SMD) may lead to a drought situation to which some growing crops are vulnerable. When applied to grass, the simulation model predicts reduced growth rates due to limiting moisture on the following basis: (a) 0% reduction up to 40 mm SMD; (b) 25% at SMD of 60 mm; (c) 50% at SMD of 80 mm and (d) 75% at SMD of 100 mm. A SMD greater than 80 mm is regarded as an indicator of drought and greatly diminished grass growth. The effect of high SMD on other crops depends on the crop and on the growth stage.

Radiation
In general it is sufficient to interpolate radiation between measuring stations by the weighted scheme but taking into account physical factors such as altitude, degree of orientation of the slope of the terrain, proximity to the sea and the topographic features, hill top, exposure or shadow effect. As there is a greater number of stations measuring sunshine than recording radiation, a mix of data sources may be used taking account errors of the techniques or formulae.

Altitudinal variations in radiation relate to increased cloudiness and relative humidity with elevation. The resulting reduction in radiation can be of the order of 5% per 100 m. In the lee of mountains and in sheltered valleys, the variation becomes more complex, and the local characteristics need special attention.

The grazing season
The length of the growing season based on soil temperature alone varies from 300 days in extreme southern areas to 240 days in the north-midlands and northeast (Keane, 1986). The utilization of grass, however, depends on the state of the land trafficability, which in turn is related to the rainfall and drainage. While the length of the grazing season depends on many other factors including elevation, based on weather factors alone, Smith (1976) suggested that the length of the grazing season in days may usefully be approximated by the formula:

$$L_{grz} = 29.3 \, T_a(^0C) - 0.1RR \, (mm) + 19.5 \qquad 3.18$$

where T_a is the local annual mean daily temperature and RR the average annual precipitation. Depending on soil type, the length of the grazing season may be shorter than the growing season by 65-75 days in the southwest of Ireland to 35-40 days in the northeast of the country. Field conditions permitting, the grazing season usually begins some three to four weeks after the start of the growing season with an increasing delay of about 4-5 days for every 100 m increase in altitude (Keane, 1992).

Derivation of daily temperature from monthly values
In modelling, there may be a need to use daily values of mean air temperature, or to compare results with the long-term average for periods not coinciding with an exact month. Thus it can be useful to interpolate day to day temperatures from the annual cycle of air temperature. One method is to assume that the monthly mean temperature coincides with the middle day of the month (i.e. 15th), and use simple linear interpolation between adjacent months to derive intermediate daily values. This scheme may be sufficient as a first approximation; however, a curvilinear interpolation may be necessary at times, e.g. during the sharp recovery of temperature in spring, February/March, or during the summer months of July/August. Thus the simple linear method can overestimate or underestimate daily values. For example, peak daily temperatures in summer are not captured because, using linear interpolation, the monthly mean for the warmest month becomes the annual maximum daily value. A

better approximation to daily values is obtained by fitting an appropriate polynomial or
sine curve to the monthly means for the station.

Recognizing the limitations of linear interpolation, Brooks and Carruthers (1953)
showed that accurate daily values can be obtained by a sine wave curve of the following
form:

$$T = \overline{T}_o + a\cos t + b\sin t + c \hspace{3cm} 3.19$$

where $\overline{T}_0$, $\overline{T}_1$ and $\overline{T}_2$ are the monthly means (said to coincide with the 15th) for each
of three consecutive months, t is the day number from the beginning of the first of the
three months and T is the daily mean to be determined. Through separate integration for
each month from $t = 0$ to $\pi/6$; $\pi/6$ to $\pi/3$; and $\pi/3$ to $\pi/2$, Brooks and Carruthers (p.272)
derived the following values for a, b and c:

$$a = 5.34\,\delta_1 - 3.39\,\delta_2$$
$$b = 5.34\,\delta_1 - 1.96\,\delta_2$$
$$c = -6.47\,\delta_1 + 3.74\,\delta_2$$
$$where \;\; \delta_1 = \overline{T}_1 - \overline{T}_0 \;\; and \;\; \delta_2 = \overline{T}_2 - \overline{T}_0 \hspace{2cm} 3.20$$

From data for the mean air temperature at Clones for May, June and July (1961-1990),
calculated from the maxima and minima, are 10.1 ^{0}C, 12.9 ^{0}C and 14.5 ^{0}C respectively.
Thus $\delta_1 = 2.8\,^0$C and $\delta_2 = 4.4\,^0$C, and a = 0.036, b = 6.328 and c=
-1.660. The mean temperature derived by formula for June 15th is 12.94 ^{0}C which
compares favourably with 12.9 ^{0}C.

3.6 New Sources of Meteorological Data

New sources of meteorological information, which complement conventional observing
systems, include data from radar, weather satellites/remote sensing and numerical
weather (analysis and) prediction (NWP), each with its own strengths and weaknesses.
While the data from these sources can often be uncertain, either in terms of over
smoothing (as in the case of NWP; *see* Chapter 8) or over/underestimation, nevertheless
the additional information can be particularly useful, especially in data-sparse or data-
void areas.

Radar
Weather radar permits precipitation to be measured continuously in space and time over
a wide area. Thus weather radars can provide rate of rainfall, e.g. mm per hour, or can
be used to accumulate rainfall over given time periods and over large areas such as river
catchment areas. While radar can identify the presence of precipitation from a great
distance, the useful range for quantifying rainfall amounts, however, is considered to be
100 km. As part of the European network of radars operated by the relevant national
meteorological services, there are three in Ireland, at Dublin and Shannon Airports, and
at Castor Bay in Northern Ireland (see Collins and Cummins, 1996). Radar data
composites for Dublin and Shannon are archived in the Met Éireann data base since

1997, and Castor Bay archived data are available through the Meteorological Office, Belfast.

Satellite/Remote Sensing Data
Considerable advances have been made in deriving geophysical products from remote sensing such as the polar-orbiting NOAA AVHRR, the European and US geostationary weather satellites, METEOSAT and GOES. The NOAA AVHRR, from which NDVI (Normalised Difference Vegetation Index) images are derived, provides semi-quantitative estimates of crop growth and yield, land surface hydrology, surface temperature, animal and plant diseases, and ozone and pollution monitoring. The main limitation in relation to Ireland with respect to NOAA is that the number of useful passes over the country is very limited because of cloud cover. Archives of radiances from EUMETSAT are maintained at the Darmstadt Centre in Germany.

The strength of METEOSAT technology lies in the depiction of precipitating clouds and the probable intensity of the associated rainfall. However, satellite data are still not sufficiently accurate to quantify precipitation amounts over short time intervals with reasonable accuracy. The introduction of pre-processing facilities, entitled Satellite Application Facilities (SAF), have advanced matters in this respect. The launch of the Second Generation METEOSAT (SGM), with radiometric imaging on 12 channels as opposed to three, and the proposed European Polar-Orbiting System (EPS), will open up new possibilities for providing greater accuracy and detail in the various products (Pankiewicz *et al*, 1998).

3.7 Conclusion

Weather reports from a network of meteorological stations do not always reflect the variety of conditions existing in a region. Thus simple extrapolation is sometimes insufficient to depict the spatial extent of many weather phenomena. Not only must users of meteorological data be aware of the quality of the data itself, but equally they must employ appropriate methods of data interpretation, extrapolation or interpolation. Consideration must be given to a change of scale from site specific input information to outputs often consisting of analyses or forecasts that are interpreted over extended regions, or *vice versa*. Progress in bringing technological developments on crop-weather simulation modelling into everyday farm management use will be greatly facilitated by having ready access to meteorological data on an operational basis. This chapter endeavours to provide users with a basic understanding of the various sources of climatic data, their availability and range of validity, so that the potential of the resource can be maximized to better inform farm management.

Methods for on-line access to meteorological databases held by National Meteorological Services are presently undergoing great change and expansion. Access to data through e-mail and a proliferation of web sites is gradually becoming more the norm. E-mail facilities are increasingly used for the movement of vast amounts of data and information to accredited users/customers, often in near real-time. The transmission of radar and satellite data, recent hourly weather reports from the network of stations, are suited to these communication networks.

References

Aslyng, H.C. (1965): *Evaporation, evaporation and water balance investigations at Copenhagen 1955-64. Acta Agriculturae Scandanavica*, Vol XV. pp 284-300.

Brereton, A.J. (1989): Analysis of the effects of water-table on a grassland farming system. In Dodd,V.A. and Grace,P.M. (Editors): *Land and water use* Proceedings 11th International Congress on Agricultural Engineering. Balkema. Rotterdam. pp 2685-2690.

Brereton, A.J., & C.J. Korte. (1997): The Definition of Agrometeorological Information Required for Pasture and Livestock production in Temperate Regions. Chapter 6: *Analysis of the effects of water-table on a grassland farming system* (pp 23-28). CAgM Report No. 71, WMO, Geneve.

Brooks, C.E.P. & N. Carruthers. (1953): *Handbook of Statistical Methods in Meteorology*. M.O. 538. Her Majesty's Stationary Office, London.

Budyko, M.I. (1974): *Climate and Life*. (English Editor: D.H. Miller), Academic Press, London.

Collins, J. F. & T. Cummins (Editors) (1996): *Agroclimatic Atlas of Ireland*. AGMET, c/o Faculty of Agriculture, University College, Dublin 4.

Connaughton, M.J. (1967): Global Solar Radiation, Potential Evapotranspiration and Potential Water Deficit in Ireland. *Technical Note No. 31*. Met Éireann, Dublin.

Cressie, N A C, (1993): *Statistics for Spherical Data*. Wiley, Chichester.

Doorenbos, J. & K.O. Pruitt. (1977): *Crop Water requirements. FAO irrigation and drainage paper 24* (revised). Rome: Food and Agriculrural Organization of the United Nations.

Fitzgerald, D. & F. Forrestal. (1996): Monthly and Annual Averages of Rainfall for Ireland 1961-1990. *Climatological Note No. 10*. Met Éireann, Dublin 9.

Guyot, G. (1998): *Physics of the Environment and Climate*. Praxis Publishing/John Wiley & Sons, Chichester, UK.

Hamilton, J.E.M., P. Lennon, B. & O'Donnell. (1987): Objective Analysis of Monthly Climatological Fields of Temperature, Sunshine, Rainfall Percentage and Rainfall Amount. *Journal of Climatology* **8**: 109-124.

HMSO. (1967): Potential Transpiration - For use in Irrigation and Hydrology in the United Kingdom and Republic of Ireland. *MAFF Technical Bulletin No. 16*. HMS Stationery Office, London.

Hooghart, J.C. (Editor). (1987): *Evaporation and Weather. Proceedings and Information No. 39*. The Netherlands Organization for Applied Scientific Research. The Hague.

Hough, M.N., R. Gommes, T.Keane, & D. Rijks. (1998): Input weather data. In: *Agrometeorological Applications for regional Crop Monitoring and Production Assessment*, Eds: D. Rijks, J.M. Terres, P.Vossen. Joint Research Centre/European Commission, Ispra, Italy.

Hough, M.N. (1998): Criteria for Use of weather Stations. In D. Rijks, J.M. Terres, P.Vossen (Editors): *Agrometeorological Applications for regional Crop Monitoring and Production Assessment*, (Annex 4A). Joint Research Centre/European Commission, Ispra, Italy.

Hough, M.N., S. G. Palmer, M. Lee, I.A. Barrie & A. Weir. (1996): *The Meteorological Office rainfall and evaporation calculation system: MORECS version 2.0 (1995)*. Meteorological Office, Bracknell.

Hough, M.N. & R.J.A. Jones. (1997): The Meteorological Office rainfall and evaporation calculation system: MORECS version 2.0 – an overview. *Hydrology and earth System Sciences* **1**: 227-239.

Iqbal, M. (1983): *An Introduction to Solar Radiation.* Academic Press, London.

Keane, D. (1985). Air Temperature in Ireland 1951-1980, monthly, seasonal and annual mean and extreme values. *Climatological Note No. 8.* Met Éireann, Dublin.

Keane, T. (1982): Weather and Potato Blight. *Agrometeorological Memorandum No. 8.* Met Éireann, Dublin.

Keane, T. (Editor). (1986): *Climate, Weather and Irish Agriculture.* AGMET Group, J.F.Collins,ᶜ/o Faculty of Agriculture University College Dublin.

Keane, T. (Editor). (1992): *Irish Farming, Weather and Environment.* AGMET, ᶜ/o Met Éireann, Dublin.

Keane, T. (1998): Interrupted Sequences of Observations. In D. Rijks, J.M. Terres, P.Vossen (Editors): *Agrometeorological Applications for regional Crop Monitoring and Production Assessment* (Annex 1A). Joint Research Centre/European Commission, Ispra, Italy.

Kitandis, P.K. (1997): *Introduction to Geostatistics: Applications to Hydrology.* Cambridge University Press.

Meteo Consult BV. (1991): *AMDaC User Manual,* Meteo Consult BV, Wageningen.

Pankiewicz, G.S., P. Butterworth, B.J. Conway, A.R. Harris, D.C. Jones, J.K. Ridley and M.A. Ringer. (1998): An Investigation into Environmental Satellite Imagery Products. *Forecasting Research Technical Report No. 248* (September). Meteorological Office, Bracknell, UK.

Perrin de Brichambaut, C. (1975): 'Cahiers A.F.E.D.E.S.', *supplément au no. 1.* Editions Européennes Thermique et Industrie, Paris.

Pereiro, A.R., N.A Villa Nova. (1992): Analysis of the Priestly-Taylor Parameter. *Agricultural and Forest Meteorology* **61**: 1-9

Rijks D, J.M. Terres, & P.Vossen. (1998): *Agrometeorological Applications for regional Crop Monitoring and Production Assessment,* Eds: D. Rijks, Joint Research Centre/European Commission, Ispra, Italy.

Rohan, P. K. (2nd Edition). (1986): *The Climate of Ireland.* Government Sales Publications. Dublin.

Smith, M., R.G. Allen, J.L. Monteith, A. Perrier, L. Pereira & A. Segesen. (1991): *Report on the Expert Consultation on Procedures for revision of FAO Guidelines for Prediction of Crop water Requirements.* FAO, Rome.

Smith, L.P. (1976): The Agricultural Climate of England and Wales. *Technical Bulletin 35.* Her Majesty's Stationary Office, London.

Villapando, J.F. (Chairman of Expert Group) (1992): Practical Use of Agrometeorological Data and Information for Planning and Operational activities in Agriculture. *CAgM Report No. 60,* WMO/TD-No. 629, WMO, Genevè,

WMO. (1990): *Guide to Instruments and Methods of Meteorological Observation* (5th edition). World Meteorological Organization, Genevè.

Chapter 4   Terrestrial Data

J.F. Collins[1], E. Daly[2], J. Sweeney[3], M. Walsh[4], J. White [5] and G. Wright[6].

1. *Department of Crop Science Horticulture & Forestry, University College Dublin, Ireland*
2. *Eugene Daly Associates, Delgany, Co. Wicklow, Ireland*
3. *Department of Geography, National University of Ireland Maynooth, Co. Kildare, Ireland*
4. *Teagasc, Athenry, Co. Galway, Ireland*
5. *Department of Botany, University College Dublin, Ireland*
6. *Geological Survey of Ireland, Beggars Bush, Dublin 4, Ireland*

4.1   Introduction

The purpose of this chapter is to record the sources of data (with examples based on Irish resources) that can help a researcher's work. Its major beneficiaries are expected to be those who assemble Geographical Information Systems and those involved in the modelling process. A selection of topics has been chosen to show the range of information that is available and where shortcomings exist. Map data acquisition and storage, modern developments in spatial representation and the use of such data in land resource appraisal are introduced. Groundwater, surface water, marine resources, soil and vegetation resources are assessed. The discussion of land information and appraisal shows how data from a wide variety of sources, collected by personnel from disparate institutions and for widely differing purposes, can be assembled, reprocessed and interpreted for a range of uses – uses, perhaps, that the original surveyors never even dreamed of!

4.2   Map data and spatial representation
(M. Walsh)

4.2.1   Introduction

The earliest cartographers first established terrestrial databases in the form of maps which enabled mankind to extend the frontiers of human habitation and to further develop areas which were already known. During the nineteenth century and particularly in the twentieth century, large amounts of data in relation to terrestrial features, climate and the oceans accumulated. The interpretation of data, especially for areas which are distant from the points of collection, presents a challenge to modern scientists. This challenge has been met by the construction of various types of databases, models and expert systems.

4.2.2 Topographic Map Data

Maps are one of the primary methods for conveying the results of field studies. Scale and legibility determine the detail at which data and spatial information are presented. The myriad applications of maps is well demonstrated in modern atlases, e.g. Royal Irish Academy's *Atlas of Ireland* (RIA, 1979). The more recent *Agroclimatic Atlas of Ireland* (Collins & Cummins, 1996) presents a selection of over 100 maps showing a wealth of Irish data relevant to agriculture and the environment. Maps containing information on topography and administrative units generally form the base for interpretation and presentation of resource data.

The Ordnance Survey (OSI, 1998) is the principal producer of topographic maps for Ireland, the most popular scales being 1:2,500, 1:10,560, 1:50,000 and 1:126,720. They are available in hard copy and in a wide variety of electronic formats. The most detailed map scale with country-wide coverage is the six-inch (1:10,560) which is now being reproduced at a scale of 1:10,000. A wealth of information is contained in these sheets in the form of lines, symbols and text. Boundaries of counties, baronies, parishes and townlands have specific line representations and appropriate combinations of these where boundaries coincide. Roads, railways, rivers and streams and associated features as well as a variety of others such as quarries, gravel pits, antiquities, wells, pumps, trigonometrical stations and contours all have specific lines, symbols or combinations of both. Symbols for vegetative cover differentiate between coniferous, broad-leaved and mixed forestry, orchard, brushwood, rough pasture, osieries, reeds and marsh. Text often accompanies these symbols, indicating the names of the administrative units and the nature of a range of public buildings. The six-inch series is intensively used in a wide range of activities involving agriculture, engineering, forestry, environment and geography.

The "25-inch" series (1:2,500 scale) covers most of the intensively-farmed areas and omits the sparsely-inhabited areas of the country. It contains most of the same information as the 6-inch series but in greater detail. The area of each field, in acres, correct to the third decimal place is shown, but contour lines are omitted. The series is regularly used by farmers, local authorities, government officials and by most agencies and institutions that deal with land and property.

Both the 6-inch and the 25-inch maps are based on the Cassini Projection which is a version (conformal) of the Transverse Macerator Projection. A separate projection, i.e. standard latitude and longitude, is used for each county in order to limit scale errors. Within this scale range are the Area Aid maps (scales from 1:2,500 to 1:10,000, depending on size of holding) which show land parcels, currently the smallest administrative unit, complete with area (hectare), townland name, and alpha-numeric code. These are very useful aids for planning nutrient management, crop rotations, recording land use history, yields and similar types of information.

A new series of 89 topographic maps (Discovery Series (Republic)/Discoverer Series (Northern Ireland)) covering the whole island at a scale of 1:50,000 (2 cm to 1 km) has been completed which replaces the earlier half-inch coverage. An index to the sheets in the series is printed on the cover of each map. The earlier sheets were compiled from

1970s aerial photography (1:30,000 nominal scale) and later sheets from 1995 aerial photography (1:40,000 nominal scale). The map legend includes a range of features under the general headings: tourist information, roads, water, railways, antiquities, relief, boundaries and general features. Contours are inserted at 10 metre intervals and colour-layered at 100 metre intervals. The original Gaelic version is included for some of the place names. The National Grid at 1 x 1 km intervals is super-imposed in blue lining on all sheets. In addition, areas of particular interest are being published at a scale of 1:25,000, based on the same aerial photography. Lists of maps and related publications are available from both the Ordnance Survey of Ireland (OSI, 1998) and the Ordnance Survey of Northern Ireland (OSNI, 1994).

The island of Ireland is usually depicted on a single sheet at scales ranging from 1:500,000 to 1:750,000. General soil, land-cover and peatland maps are at 1:575,000; the 1997 edition of the O.S. Road Map is scaled to 1:600,000; administrative areas, index maps to 6-inch sheets and aerial photographs as well as the catchment basin map are depicted at "ten statute miles to 1 inch" (1:633,600) while the popular geological map was published at the slightly smaller scale of 1:750,000. The scales used in textbook-sized atlases (e.g. Aalen *et al.* 1997, Collins & Cummins 1996, Horner *et al.* 1984, Lafferty *et al.* 1999) range from c. 1:2,000,000 to 1:20,000,000.

4.2.3   Geo-referencing data and satellite positioning systems

In recent decades the establishment of accurate location, whether according to latitude, longitude or national grid co-ordinates is being overtaken by a combination of modern communication technology, photogrammetry, and both geostationary and orbiting satellites. There are currently two active positioning systems (the U.S.-administered Global Positioning System (GPS) and the Russian Federation Global Navigation System (GLONASS)), each with a world-wide coverage but separate constellations of satellites. A third system, EUSAT, is being developed by EU member states. Based on signals received at ground level from a number of these satellites the position of any point on the earth's surface can be established. The accuracy depends on the sophistication of the receiver, time of exposure, and number of satellites within range. The accuracy range for civilian uses is c. 10 m. Use of differential correction (DGPS) can improve this value to <1 m. Post-processed survey grade, carrier-phase receivers can achieve a static and dynamic accuracy measured in millimetres.

The software incorporated in GPS receivers is programmed so that the readout refers to the local National Grid system and hence longitude as the Easting, or x co-ordinate, latitude as the Northing, or y co-ordinate, and altitude as the z co-ordinate of the local system. The Irish national reference point is an OSI station at Malin Head (OSI, 1996). An example of GPS data for positioning is given in Table 4.1. It shows Easting, Northing and altitude co-ordinates for selected observation points on the Teagasc Hill Sheep Farm, near Leenaun Co. Mayo (Egan *et al.*, 1996). A control point (concealed concrete marker slab) which acts as a "base station" was established on the farm by DGPS, using the known co-ordinates of a nearby trigonometrical station.

A simple practical use of GPS is to locate a point in a landscape devoid of map reference features such as buildings, fences, rock outcrops or even large trees. Examples of such landscapes include large tracts of moorland, blanket or raised bog, and lakes. Large uniform fields (20 ha upwards) may be included. This technology has made the accurate mapping of natural resources possible. Problems posed by forest and urban areas as well as by electromagnetic interference are being researched. GPS is the standard georeferencing device in aircraft used for data capture and is being installed in equipment used in precision agriculture. Good quality GPS equipment can be purchased or hired from a number of specialist firms dealing with surveying. Data on GPS reference points, trigonometrical stations, and related services can be obtained from the Ordnance Survey Office.

Table 4.1: Some georeferenced data for the environmental monitoring site, Teagasc Hill Sheep Farm, Leenaun, Co. Mayo

Site Identification	Easting (m)	Northing (m)	Altitude (m)
Control	092,538.014	266,090.287	39.103
Peg L15	091,886.255	266,034.318	32.103
Peg F21	092,190.183	266,730.985	126.327
Peg H31	093,166.473	267,028.157	70.130
Peg C10	091,077.944	266,530.427	259.507

4.2.4   Aerial Photogrammetry

Modern map making is based on aerial photographs. Photographic film is exposed in a camera fixed to the underside of specially equipped planes and flown at a selected altitude under cloud-free conditions. Modern aircraft can carry a variety of sensors including conventional aerial cameras, small (35mm) cameras, video recorders, digital sensors, thermal cameras and radar sensors. Spatial resolution can range from a few centimetres to several metres.

Panchromatic aerial photography giving cloud-free stereoscopic coverage was obtained for the whole of Ireland in 1974-1977 at a nominal scale of 1:30,000 and in 1995 at a scale of 1:40,000. Most of the west coast of Ireland has similar coverage dating from the late 1940s to the early 1950s. More limited coverage, mainly for planning and environmental activities, has been completed for a number of areas. Ground truthing is essential to interpretation in all cases, and accurate positioning is recommended for geo-referenced ground control. A scanning and rectification procedure using computer hardware and software is now used to correct photographic distortion. (Developments in digital camera technology are replacing the need for scanning). The end-product, which is known as an orthophoto (or orthophotomap), is a medium suitable for many purposes such as plotting, mapping and measuring.

The 1995 aerial photography, at 1:40,000, has been used to compile the Area Aid maps which show individual land parcels. With adequate ground control, it can provide measurements of individual objects (buildings, trees, channels) to sub-metre accuracy. Photographic products also have specialised applications e.g. engineering, road planning, and monitoring land use changes. Photo-products can be purchased from commercial companies and from institutions such as GSI and OSI. A number of companies provide flying services, production of diapositives and prints, and facilities for scanning.

4.2.5   Surface Modelling

The data captured by aerial photography (and some satellite imagery) can be stored and processed by computer technology which manipulates and processes data describing the earth's surface. When an "actual" land (or sea-bed) surface is so created the product is referred to as Digital Elevation Model (DEM) or Digital Terrain Model (DTM). The data components used for modelling such surfaces are generally either: (1) *Grid Structure/Altitude Matrix*: the height of the surface is estimated on a regular grid basis over the area concerned; (2) *Contour Slice*: height data are provided by contour interval, e.g., 50m 100m. The product is essentially a thematic map which may be either in vector or raster format; (3) *Triangular Irregular Network (TIN)*: height data are chosen at random but increase in density with increase in complexity of the topography. They are then linked to form triangular facets. The TIN is generally a vector-based representation of a surface; or (4) *Digital Elevation Models (DEMs)*: help to highlight anomalous values and spurious patterns. They have been very useful in the study of hydrology, soil, topography, climate, interpretation of stereo models, interpretation of satellite imagery and visual-effect-simulated perspective views.

This technology improves the presentation of map data. The data can be enhanced in appearance by presenting them as a "shaded relief map" or a "simulated perspective view" (Figure 4.1). Different perspectives can be had by rotating the image so that the viewer can see it form any chosen point. Contour (10 m) and DTM (10 m grid) information, derived either from the 1974-1977 or from 1995 photography, is available in 20 x 20 km tiles, under a lease-purchase arrangement from the OSI.

Figure 4.1: Draped perspective image of a portion of Derreens Hill, Co. Mayo

Maps are representations of reality, reduced in size. A well-made map is the product of a sophisticated range of procedures which may include recording, calculating, analysing, displaying, organising, and presenting, on paper or screen, the spatial relationships between things. Since maps are a system of communication in the same way as written and spoken languages, they demand that users are acquainted with graphicacy and the science of cartography. In the sections that follow there are many examples where familiarity with maps, their contents, usefulness and hidden limitations will become apparent to the reader. This understanding is necessary for reliable application of GIS technology (see Chapter 2).

4.3 *Ground Water Resources*

(G.R. Wright)

4.3.1 Introduction

Groundwater is generally taken to encompass all water below the water table. It is sometimes also taken to include water within the capillary fringe above the water table. Groundwater is stored within, and moves through, the spaces of granular deposits, and the fissures (fractures, joints) of rocks.

Groundwater contributes 20-25% of drinking water in Ireland, compared with about 33% in England and Wales, 3% in Scotland, and 7% in Northern Ireland. The percentage contribution in some European mainland countries is much higher; in Austria it reaches 99%. Although our major cities are supplied almost exclusively from surface water sources, many towns, villages and factories, and innumerable private homes and farms, depend on groundwater.

Fissure-flow aquifers predominate in Ireland. Aquifers with intergranular flow are restricted to scattered Quaternary sand and gravel deposits, which are locally important but rarely extensive enough to be major aquifers. The main fissure-flow aquifers are: (1) *Carboniferous limestones*, which are very widespread but form major aquifers only where they are sufficiently clean and fissured or dolomitised. In some regions, notably the north-west, west and south, the limestones are extensively karstified; (2) *Devonian, Carboniferous and Permo-Triassic sandstones*, some of which also have some degree of intergranular flow; (3) *Cainozoic fractured basalts* in north-east Ireland; (4) *Cretaceous chalk* in north-east Ireland; and (5) *Ordovician volcanic rocks*, strongly fractured, which form a narrow NE-SW-trending belt in south-eastern Ireland (mainly counties Wexford and Waterford).

Irish aquifers are predominantly shallow, so few wells are deeper than 120 metres, and the water table is generally high, thanks to abundant rainfall. Most aquifers are unconfined, although local confinement by glacial till is fairly common. These factors mean that many aquifers are highly or extremely vulnerable to pollution.

4.3.2   Data Sources

Tabular Data

The Geological Survey of Ireland (GSI) is the main repository of groundwater data in Ireland, but the Environmental Protection Agency (EPA) also collects water level and water quality data. The Geological Survey of Northern Ireland (GSNI) performs an equivalent function in its region. The GSI's groundwater data have accumulated over a period of about thirty years, and for the most part are contained in four principal databanks: (1) wells and pumping tests; (2) chemical analyses of groundwater samples; (3) groundwater level monitoring; and (4) karst features. GSI also has an extensive and growing collection of reports on hydrogeology and related topics. The EPA (which has a statutory responsibility to collect hydrometric and water quality data, including data on groundwater) has established a national monitoring network for groundwater levels and groundwater quality with some of the data now published.

As in many countries, groundwater data in Ireland are very patchy. In any given area, good data exist for only a few wells or springs where intensive investigations have taken place, but there are many wells and springs for which few data are available and probably many others which are essentially unrecorded. This situation has several causes, primarily the absence of any statutory reporting of well drilling, the general lack of concern about groundwater resources (at least until recently), and the lack of investment in groundwater studies. In this situation, groundwater investigators (needing answers in a relatively short time) must make the best of patchy data, resisting the temptations of, on the one hand, dismissing poor data as useless, and on the other, of over extrapolating good data. European groundwater legislation will be rapidly changing this in the years to come.

Well Data

About 30,000 records of water wells in the GSI data banks have been supplied on a voluntary basis, but may represent only 10-20% of the actual total in the country. Well data have been derived in different ways; by well drillers, local authorities, consultants, and by field surveys. In each case the type of information on record differs; for instance, records submitted soon after drilling may include some details of the geological formations penetrated, whereas field surveys do not usually unearth this information. On the other hand, a field survey provides a precise location and an accurate water level, often lacking in other records. Some details of pumping tests are available for about 1,200 wells. Of the GSI's well databank, about 25,000 are basic records with minimal data (generally lacking precise locations, geological or yield data), and about 5000 have more detailed information. In addition, detailed well surveys have been carried out in a few places. These approach 100% coverage of wells in their areas and achieve precise locations, but still record rather minimal data; geology or yields are rarely obtainable. The records are filed according to counties and by Ordnance Survey 6-inch sheet numbers. So far, all available groundwater data (30,000+ records) have been computerised for 13 counties where groundwater protection schemes have been undertaken. The well records are used for aquifer classification, drawing water table or piezometric maps, constructing depth-to-bedrock maps, and providing an initial list of possible sites for further data collection.

Hydrochemical and Water Quality Data
The GSI hydrochemical databank comprises some 3000 fairly complete groundwater analyses, plus a roughly equal number of partial chemical and bacteriological analyses. The records are filed according to counties and by Ordnance Survey 6" sheet numbers. Like the well records, these analyses are patchy; some are of doubtful validity (lacking ionic balance), some are not precisely georeferenced and some lack any details of the water source. In a few cases there are enough repeated analyses from a given source to examine changes in water quality over time. For counties where GSI has undertaken groundwater protection schemes, the hydrochemical data are entered into a computer database The analyses for the remainder of the country still await study, evaluation and processing.

The EPA has initiated a nation-wide network to monitor groundwater chemistry and quality, and as these data accumulate they will constitute the country's main groundwater quality database that might be supplemented by additional local authority data. Groundwater quality and chemistry data can be used to characterise the baseline water quality in aquifers, to define trends in water composition through space and time, to deduce the underground pathway which water has followed and to infer the vulnerability of aquifers

Groundwater Level Data
The GSI has a unique data collection recording groundwater level fluctuations in about 40 wells in seven counties. These records have been computerised and provide a good, though discontinuous, picture of groundwater levels in Ireland over the past 30 years, and yield valuable information about the nature of the aquifers and the groundwater regime. Groundwater level monitoring data are essential for: (1) calibrating many predictive groundwater models, which are increasingly required for Environmental Impact Assessments; (2) assessing the feasibility of landfills, quarries, septic tanks and other effluent disposal, excavations for roads and building foundations and amelioration of flooding problems; (3) characterising aquifer recharge and vulnerability to pollution; (4) observing major groundwater abstractions; (5) determine the minimum depths of wells; (6) measuring the effects of changes in surface water abstraction; and (7) are essential for water balance studies.

To date, long-term groundwater level monitoring data are available only for limited areas of Ireland, mainly from GSI work. A new national monitoring network has recently been established by the EPA, with involvement by GSI, Office of Public Works (OPW) and local authorities. This groundwater level network will complement the existing long-established networks for rainfall, evaporation and surface water flow, the other main components of the hydrological cycle.

4.3.3   Reports and Maps

Groundwater Reports
The GSI has a large number of reports, ranging in size from a few pages to two or three volumes, and in subject matter from very local issues of water supply or pollution to regional and national summaries. They are available on request.

National scale maps

The early 1970s saw the first serious attempts to compile aquifer maps of Ireland. Three versions were produced: (1) a 1:1.5 million, for the International Association of Hydrogeologists (IAH) International Hydrogeological Map of Europe. This was compiled around 1972 and published in 1976 and 1980, with explanatory memoirs published in 1978 and1980. The Northern Ireland portion of the map was compiled by GSNI; (2) a 1:2 million for the Royal Irish Academy's *Atlas of Ireland* (RIA, 1975); and (3) a simplified version at 1:2 million, showing rock aquifers only (omitting sands and gravels), reproduced in *Mining Ireland* (Aldwell, 1975), in *Technology Ireland* (Wright, 1976) and in *Water Wells* (Anon., 1977). The usefulness of these maps was limited by their small scale. In 1979 GSI produced a more comprehensive national aquifer map at 1:500,000 as part of a European Commission-sponsored project. This map was published (on parts of three sheets) in 1982 with an explanatory report. Again, the Northern Ireland portion of the map was compiled by GSNI. The aquifer classification system used was chosen specifically for the project, and ignored many minor aquifers. Additional maps showing hydrological data, groundwater abstractions and surplus groundwater resources were also published. A small scale (1:1.5 million) version of the map (with minor revisions) was produced by GSI, GSNI and K.T. Cullen & Co. in 1995.

A follow-up project sponsored by the EC produced a national map of Groundwater Vulnerability at 1:500,000, compiled in 1983 but not published. This was again accompanied by a report and a series of A4-size 1:500,000 maps of groundwater quality as represented by Hardness, Chloride, Total Dissolved Solids and "Excess Substances". These quality maps only dealt with the aquifers as designated by the earlier study. The Vulnerability definition used in this project depended on the vertical "Time of Travel" for infiltration from the ground surface to the aquifer or water table, and was different from that currently used in GSI.

A new national aquifer map, covering all types of aquifers, is planned by GSI at a scale of 1:625,000. GSNI has produced an aquifer map and groundwater vulnerability map of Northern Ireland at 1:250,000 where the definition of groundwater vulnerability used is the same as used in Britain but different from that used by GSI in Ireland.

Medium-scale maps (1:100,000 to 1:50,000)

Examples include: (1) the north-east Regional Development Organisation maps; (2) the Nore River Basin and associated maps; (3) Groundwater Protection Scheme maps currently available (at 1:50,000 or 1:63,360) for counties Claire, Cork (south), Laois, Limerick, Meath, Offaly, South Tipperary, Waterford, and Wicklow (draft). Each suite includes maps of bedrock geology, subsoils (Quaternary geology), depth-to-bedrock, hydrogeological data, aquifers, groundwater vulnerability, and groundwater protection zones. An earlier suite (1979) at 1:63,360 scale is available for County Dublin (bedrock geology, Quaternary Geology and Aquifers only); and (4) maps by consultants for various local areas.

Large Scale Maps (1:25,000-1:10,000 and larger)
GSI has produced maps in this scale range for Source Protection Areas around some major public groundwater sources. Other hydrogeological maps at these scales have been produced by consultants for various local areas and regions. Detailed information is available on request.

4.4   Surface Water Resources

(E. Daly)

4.4.1   Introduction

Surface water in the form of lakes and rivers is the most visible part of the hydrologic cycle on the global landmass. Here water is concentrated into relatively small areas of the earth's surface. In Ireland there are about 16,000 km of major and minor river channels and around 4,500 lakes and large ponds which occupy approximately 2% of the island's landmass.

Surface waters have numerous beneficial uses for the human population, such as water supply, waste attenuation, recreation and transport. There is a very definite seasonal aspect to water in this form. In Ireland much of the water in streams and rivers in the winter is derived from relatively recent rainfall whereas in summer the flow in rivers is maintained from drawdown of storage in geological strata. Extreme surface water levels often have a significant impact on the human consciousness. The floods of 1954, 1968, 1978, 1986 (Hurricane Charlie), 1989-'91 and 1995 (Gort) and droughts of 1959, 1975-1976 and 1995 are all remembered.

The natural chemical quality of surface waters is a function of the flow regime, the geology of the catchment area and the time of year. Water quality is also influenced by land use and the degree of human activity within the upstream catchment. Surface water bodies provide a rich habitat for fauna and flora; type and diversity are indicative of the health of waters. Wetlands which have characteristics half-way between land and surface water are often the last remaining remnants of the native environment in an area.

The use of surface water flow and quality data is now an integral part of the planning and regulatory process for most large developments and for the licensing of existing ones. Many industrial and commercial enterprises are required to monitor the receiving water quality into which they discharge treated effluents. Since the passing of Ireland's first Water Pollution Act, in 1977, many EU Directives, laws, statutory instruments, standards and guidelines have been enacted. A similar situation exists in Northern Ireland. Since the early 1980s water quality management plans or strategies have been prepared for many of the major catchments in the two jurisdictions. Ireland is divided into 40 hydrometric areas (Figure 4.2) each of which comprises a single large river catchment or a group of smaller ones. They are grouped into seven Water Resource Regions that are of such size that the water requirements of each region can be supplied from the available resources within that region (Department of Local Government,

1974). A considerable volume of data are available on Irish surface waters in both Ireland and Northern Ireland including flow, chemistry and biology. However, only some of the data, especially that of water quality, are available in a readily accessible form.

4.4.2   Tabular Data

Rivers and lakes in Ireland are regularly monitored at numerous locations for water level (stage), chemistry and biology. The basic water level data are subsequently manipulated to provide flow rates and other statistical information. Much of the chemical and biological data are concerned with water quality and the degree of deterioration relative to the natural state. These data are aggregated into chemical and biological quality ratings.

Figure 4.2: Hydrometric areas of Ireland

On a national scale surface water data (water level/stage) are collected mainly by the Rivers Agency (an agency of the Department of Agriculture) in Northern Ireland and by the Hydrometric Section of the Office of Public Works (OPW) in Ireland. Both the Environment and Heritage Service (EHS, an agency within the Department of the Environment for Northern Ireland) and the Environmental Protection Agency (EPA, formerly An Foras Forbartha [AFF] and the Environmental Research Unit [ERU]) in Ireland carry out extensive programmes for the collection of chemical and biological data. There are also a number of other organisations active in the collection of surface water data, but on a more local basis, or on specific aspects of surface water hydrology. Such organisations include the local authorities, the Electricity Supply Board, Fisheries Boards, third level colleges and some commercial organisations. Most of the data are contained within large databases operated by the OPW and EPA in Ireland and the Rivers Agency and EHS in Northern Ireland. The OPW and Rivers Agency supply flow data for particular stations on request. The EPA and its predecessors have regularly published both river flow and quality data on a national basis over the last 25 years. Since the 1980s, the EHS has published the results of its water quality monitoring programme in reports at regular intervals. Data are also available in less formal databases in other organisations, especially the local authorities.

Water Flow

Systematic river flow recording began in Ireland in 1939. Automatic water level recorders were installed at locations on many rivers in the 1940s and 1950s. In many cases the site-selection and monitoring objective was to record high (flood) flows and aid drainage projects which resulted from the passage of the Arterial Drainage Acts in 1945. The Electricity Supply Board began river gauging in the late 1920s and a network of gauges was set up in the 1930s on rivers that had some potential for hydroelectric development. More recording stations were set up in the 1970s. Most of these stations were set up to record low flows with a view to ascertaining the assimilative capacity of streams and rivers for pollution studies and water quality management plans.

The 1,369 water level gauging stations in Ireland are located on rivers, lakes and at the coast. Of these, 539 are gauging stations fitted with automatic recorders (MacCarthaigh, 1999). A register of all hydrometric gauging stations in Ireland maintained by the EPA contains details of the type of installation and the length of the flow record. Automatic water levels recorders (now being upgraded to data loggers) have been in operation in Northern Ireland since 1970, where there are now over 100 automatic water level recorders. Seventy are used to monitor flow and the remainder are used to record water levels.

Types of data.

River flow data are initially collected as a water level (stage height) above a datum. It may either be in the form of a single value (graduated staff gauge) at a particular time or a continuous chart (autographic stage recorder). Each gauging station is calibrated by taking the results of a series of river flow measurements (with a current-meter) at different stage levels to compile a rating curve of water level versus river flow (stage-discharge relationship). The rating curve is then used to compute a discharge value for a particular stage or a whole series of values (discharge hydrograph) by digitising a continuous water level chart (stage hydrograph).

The Hydrometric Section of the Office of Public Works digitises the basic water level data recorded on continuous charts (more recently the use of data loggers permits the transfer of water level information directly to a computer) and with the aid of a rating curve computes the daily mean flows. Hydrographs of daily mean flows are also available. For a given period of years additional statistical information, such as flow duration curves, reservoir storage and sustained low flows are also available. The Rivers Agency in Northern Ireland provides similar information for its gauging stations. This information is available from both organisations on request.

Data availability

Data are available in a number of different formats and publication, the details of which are outlined briefly below. Although a considerable amount of water flow and level data has been collected since monitoring began not all have been digitised nor are available in a readily readable form. The data for over 125 automatic recording stations in Ireland were published by An Foras Forbartha in Yearbooks for 1975, 1976 and 1977. In 1977 An Foras Forbartha published seven reports containing a summary of hydrometric records for each of the seven water resource regions in Ireland. These publications make readily available the results of flow measurements taken at various locations throughout the country. An Foras Forbartha (1984a) and the Environmental Resource Unit (1989) published reports on statistical analyses of river flows in three of the water resource regions. These publications are designed to provide details of the magnitude and frequency of occurrence of river flows (mainly low flows). In 1995 the Environmental Protection Agency published a text (MacCarthaigh, 1995) containing details of hydrometric stations (329, fitted with autographic recorders) and summary water balance and flow statistics. These stations are on natural rivers, i.e. rivers that are not affected by major water storage or river-flow regulation. Two reports have been published on the droughts in 1984 (An Foras Forbartha, 1984b) and 1995 (MacCarthaigh, 1995). These reports contain low-flow measurements at selected stations recorded in those years and also provide comparisons with other dry years. There is also a considerable amount of flow data, especially low-flow data for the smaller rivers, in local authority databanks. There are likely to be some low flow measurements available for most streams/rivers that receive a discharge from a local authority wastewater treatment plant. Flow data collected in Northern Ireland are sent to the National Water Archive maintained by the Institute of Hydrology at Wallingford in the United Kingdom. The data from Northern Ireland are included in annual publications (now digital) produced by the Water Archive. Data for specific sites can be obtained by e-mailing requests.

Characteristics of River Flows
Researchers in the EPA and its predecessors have compiled summary surface water statistics for the water resource regions in Ireland. The information is adapted for Table 4.2. Flow data are often required for ungauged river sections or small streams. Estimates of the flow in streams/rivers in Ireland, for which no data are available, can be obtained using water balance data, formulae and constants that are available in a number of publications. MacCarthaigh (1995) gives the long-term average runoff values calculated from rainfall and evapotranspiration data for 329 selected stations throughout Ireland. A publication by the Department of Industry and Energy (c. 1986) contains a formula for estimating the daily mean flow (DMF in $m^3\ s^{-1}$), of a river/stream at a particular location, from the catchment area and average annual runoff. Martin & Cunnane (1977; 1994) and Martin (1992) provide methods for determining dry weather flow (DWF) based on runoff.

Table 4.2: Distribution of Surface Water Resources, Republic of Ireland

Water Resource Region	Area km^2	Precipitation 1931-1960 mm yr^{-1}	Average runoff m^3 s^{-1}	Specific runoff l s^{-1} km^{-2}	Low-flow runoff m^3 s^{-1}	Specific low-flow L s^{-1} km^{-2}
Eastern	7,878	954	127.6	16.2	7.3	0.9
south-eastern	12,710	1,009	209.2	16.5	20.9	1.6
Southern	11,920	1,383	283.0	23.7	20.4	1.7
Shannon	10,830	1,003	211.5	19.5	12.1	1.1
Mid-Western	7,640	1,141	101.9	13.4	2.5	0.3
Western	8,560	1,205	292.0	34.1	10.9	1.3
North-western	9,350	1,281	262.6	28.1	8.7	0.9
Ireland	68,888		1,487.8		82.8	

4.4.3 Data Uses

Surface water flow data are used for a variety of purposes such as, water abstraction, pollution control, designs for bridges and drainage works, flood prevention and alleviation, fishery management and amenity. One example described in the literature is the use of water level data to develop a flood warning system for Kilkenny City (Shine, 1987) on the River Nore. Analysis of the historical water level charts showed that there exists a definite relationship between the flood peaks at the station at Dinin Bridge (River Dinin), some 9 km upstream of Kilkenny, and the station at John's Bridge in the City. The pre-flood level of the River Nore in Kilkenny and the rise of the level of the River Dinin at Dinin Bridge are used to predict the rise in the water level in the City with a reasonable degree of accuracy (Shine, 1987). The system gives about four hours notice of flooding which is sufficient time for the emergency services and affected population to take action to limit the worst effects.

Water Quality
Water quality of rivers and lakes in Ireland has been monitored since the early 1970s. The results of the initial surveys were published by Flanagan & Toner in 1972 (rivers) and 1975 (lakes). The work was initiated by An Foras Forbartha and is continued by the Environmental Protection Agency. Since the late 1970s reports have been compiled every four to five years on surveys of river water quality carried out over a four-year period. The initial survey assessed the quality of the 121 major rivers with catchments generally in excess of 130 km^2 and covering some 2,700km of river channel. This was subsequently extended to 7,000 km in the 1982-1986 survey. The report of the latest survey period (1995-1997) covers about 13,200 km of river channel (EPA, 1999).

The original national lake survey (1973-1974) assessed the trophic status of 53 of the larger and more important lakes (McCumiskey, 1982). In the mid-1980s the survey had been extended to 90 lakes and to 135 in the most recent period (1995-1997). The water quality survey reported in 1986 included information on 19 estuarine and coastal areas. This has been extended to 26 areas reported in 1999.

River water quality is monitored at almost 300 stations in Northern Ireland (Environment and Heritage Service, 1996). Chemical monitoring increased from a total channel length of 1,685 km in 1991 to 2,353 km in 1995 and biological monitoring increased from 2,190 km to 2,331 km over the same period. There are also over 30 monitoring stations on estuaries throughout Northern Ireland (Environment Service, 1996).

Types of Data
Water quality assessment of surface waters is based on data collected from physico-chemical and biological surveys. The two methods complement each other and provide a more detailed and balanced picture of water quality than either one alone (McCumiskey, 1991). Sampling involves both river water and benthic substrate (sediment) in contact with the water. In Ireland, river water sampling is carried on throughout the year whereas the biological surveys are normally carried out between June and October.

River water samples are generally analysed for conductivity, pH, colour, alkalinity, hardness, dissolved oxygen, biochemical oxygen demand (BOD), ammonia, chloride, ortho-phosphate, oxidised nitrogen and temperature. In addition, lake water samples are also analysed for chlorophyll, transparency, and total phosphorus. The biological monitoring of rivers is based on the relationship between water quality and the relative abundance and composition of the macro-invertebrate communities in the sediment of rivers and streams. The macro-invertebrates include the aquatic stages of insects, shrimps, snails and bivalves, worms and leeches. The greater the diversity the better the water quality. The biological information is condensed to a 5 point numerical scale (biotic index or Q values), an arbitrary system in which community composition and water quality (1 = bad to 5 = good) are related (Table 4.3). The five grades used in the general assessment of river water quality have been grouped into four classes based on the water's suitability for beneficial uses (abstraction, fishery potential and amenity value). Water Quality Index (WQI) is used to simplify the large quantity of physico-chemical data and present it in a condensed form. The trophic status of lakes is classified according to a modified version of the OECD (1982) scheme based on values of annual maximum chlorophyll concentration.

In Northern Ireland the rivers are monitored for chemistry either fortnightly or monthly and for biology three times per year (spring, summer and autumn). Water quality of rivers within Northern Ireland and Ireland are assessed using different classification systems (Environment and Heritage Service, 1996). In Northern Ireland separate chemical and biological General Quality Assessment (GQA) classification schemes subdivide water quality into six bands (Table 4.4). For cross-border studies a classification system based on the systems used in both jurisdiction, including both chemical and biological aspects, was developed (Kirk McClure Morton, 1997).

Table 4.3: Relationships of water quality to the composition of macro-invertebrate fauna (McCumiskey, 1991)

| Macroinvertebrate fauna | Water quality | | | | |
	Good (Q5)	Fair (Q4)	Doubtful (Q3)	Poor (Q2)	Bad (Q1)
Sensitive forms (A1)	+++	+	-	-	-
Sensitive forms (A2)	+++	++	-	-	-
Less sensitive forms (B)	++++	++++	+++	-	-
Tolerant forms (C)	+	++	+++	++++	?
Most tolerant forms (D)	+	+	+	++	++++

(key to Table 4.3)
++++ Abundant; +++ Common; ++ Present; +Sparse or absent; - Absent
A1   Plecoptera (excluding Leuctra), Ecdyonuridae, Ephemeridae
A2   Ephemeroptera (excluding Baetis rhodani, Cloeon, Caenis, Ephemerella)
B   Leuctra, Baetis rhodani, Cloeon, Caenis, Emphemerella, Gammarus, uncased Trichoptera, Elminthidae larvae
C   Chironomidae (excluding Chironomus), Hirudinea, Mollusca (excluding Physa)
D   Chironomus, Physa, Eristalis, Tubificidae and other Oligochaeta.

Table 4.4: Likely Uses and characteristics of classified waters in Northern Ireland (EHS, 1997)

Chemical class	Likely uses and characteristics[1]
A (very good)	All abstractions
	Very good salmonid fisheries
	Cyprinid fisheries
	Natural ecosystems
B (good)	All abstractions
	Salmonid fisheries
	Cyprinid fisheries
	Ecosystem at or close to natural
C (fairly good)	Potable supply after advanced treatment
	Other abstractions
	Good cyprinid fisheries
	A natural ecosystem, or one corresponding to a good cyprinid fishery
D (fair)	Potable supply after advanced treatment
	Other abstractions
	Fair cyprinid fisheries
	Impacted ecosystem
E (poor)	Low grade abstraction for industry
	Fish absent, sporadically present, vulnerable to pollution[2]
	Impoverished ecosystem
F (bad)	Very polluted rivers which may cause nuisance
	Severely restricted ecosystem

1. Provided other standards are met.
2. Where the Class is caused by discharges of organic pollution.

The chemical GQA system uses three variables, ammonia, biochemical oxygen demand and dissolved oxygen to classify river reaches. The biological GQA system uses a computer model called RIVPACS (River Invertebrate Prediction and Classification System) which predicts the macro-invertebrate fauna that should be present at a site in the absence of pollution or environmental stress. Comparison of the predicted communities with the observed during sampling and analysis permits the calculation of ecological quality indices (EQIs). Full explanations of the various indices and classifications used are provided in a number of Environmental Protection Agency and Environment and Heritage Service publications.

Data availability
There is a large body of water quality data available on Irish rivers, lakes and estuaries. On a national scale the information is readily accessible in the national reviews of surface water quality undertaken by An Foras Forbartha (1972, 1974, 1975 (lakes), 1980, 1982 and 1986), by the Environmental Resource Unit (1992 (lakes) and 1992) and by the Environmental Protection Agency (EPA, 1995, 1996 and 1999). The data for the two most recent survey periods (1991-1994 and 1995-1997) are available on disc. The reports contain tables that show the channel length of individual rivers in each of the four quality classes. They include text describing trends in water quality and sources of any pollution detected. The reports provide the biological quality ratings (Q values), both current and historical for the sampling stations on individual rivers. The minimum,

median and maximum values for the physico-chemical parameters and water quality indices at each sampling station are also given. Tables which contain information on lake waters and the trophic status of individual lakes are also included in these reports.

The Northern Ireland data derived from monitoring, are used to categorise river water quality into a number of classes (Table 4.4). Since the 1980s the results of river quality surveys undertaken at five-yearly intervals (EHS, 1994 and 1996) have been published. The reports contain tables which show the channel length of the seven main river systems in each of the six chemical and biological GQA classes. There are tables providing details of the numbers of private sector effluent discharges and sewage treatment works. The reports include text describing water quality in individual rivers and sources of any pollution detected. The EHS also publishes the results of investigations into particular aspects of surface water quality. Reports on Water Quality Management Plans and Strategy have been prepared for a number of large river catchments in both Ireland and Northern Ireland, which contain additional data and analysis, e.g. the Foyle (Kirk McClure Morton, 1997).

Additional information and data can be found in databases and reports in third level colleges, local authorities, fisheries boards and commercial enterprises that have to collect surface water data for planning applications, or to meet the terms of Integrated Pollution Control and waste or effluent discharge licences. The local authorities normally collect water samples upstream and downstream of their main wastewater treatment plants to determine the impact on receiving waters. They often sample streams and rivers downstream of licensed effluent discharges to monitor compliance. The fisheries boards sample surface waters with a view to prosecuting activities possibly causing pollution.

Trends in Surface Water Quality
McCumiskey (1991) compared river water quality from the initial survey in 1971 and the position in the same length of channel 20 years later. He found that both the length of seriously polluted channel and unpolluted channel had been significantly reduced. However, the corollary is that there has been a significant increase in the channel length that is both slightly and moderately polluted. McCumiskey (1991) concludes that "the main quality trend in Irish rivers over the last two decades has been a significant reduction in serious pollution and an increasing incidence of eutrophication". This trend has continued to the present (EPA, 1999). Similarly, the report on River Quality in Northern Ireland (circa 1996) noted "a decline in chemical quality, which is attributed largely to excessive nutrient enrichment (eutrophication), was evident in a number of rivers".

Data Uses
Surface water quality data has a wide variety of uses. For example it provides a baseline against which to measure subsequent improvement or deterioration in water quality, a measure of the success or failure of conservation and remedial measures, and an input into water quality management plans and calculations to assess the impact of discharges on receiving waters.

Maps
There are few national scale maps containing surface water information. The Ordnance Survey of Ireland published a map (OSI, 1958) showing the catchment areas of over 400 river catchments and coastal areas in the island. The map (scale 1:633,600) includes details of the length, in miles, of the main river channels and the altitude, in feet, of the highest ground near the source. The EPA (1995) published a map (scale 1:440,000) showing the hydrometric network in Ireland, and in 1994 it published a map at the same scale: "Ireland, River Quality 1991-1994". This map classifies the river water quality at the national survey sampling stations. Maps (scales 1:555,000) of river quality, chemical GQA and biological GQA are contained in the River Quality in Northern Ireland, 1995, report.

4.5   Marine Resources

(J. Sweeney)

4.5.1   Introduction

Ireland's continental shelf extends approximately 350 km offshore to the west, north and south of the island, consisting of mainly Palaeozoic rocks located on the slowly eastward moving edge of the European plate. More recent Tertiary-period tectonic spreading has created the Irish Sea and Celtic Sea basins to the east which are much shallower features, usually less than 50 m in depth. Ireland's marine territory encompasses some 900,000 km^2, about nine times the size of the island. This provides it with extensive marine resources in such categories as fisheries, minerals and aquaculture These resources are subject to increasing development pressures as their economic value is realised and new technologies appear to exploit their potential. Jurisdiction of the seas around Ireland has been a subject of contention since independence, particularly where unexploited energy resources or fish stocks may exist. While the maritime boundary in the Irish Sea and Celtic Sea can be relatively easily demarcated, the areas to the west are much more problematic. Competing claims from Denmark, Iceland, Ireland and the United Kingdom currently exist for parts of the Rockall Bank, an area thought to hold potential seabed wealth.

Whether the surrounding seas are considered to be a barrier, buffer, boundary or link to accessing these resources, an understanding of their potential for employment creation is important. More than any other EU country, Ireland's population consists predominantly of coastal dwellers. Over half live on the coast and some 86% live within 50 km of it. To maximise the potential benefits, and minimise the potential risks offered by this environment, good data are needed for management purposes.

4.5.2   Marine Meteorological and Climatological Data Sources

Ireland has benefited from a relatively good supply of marine meteorological data from the seas around. Fixed sources such as weather ships, manned lighthouses and light vessels have, however, dwindled as satellite platforms have increasingly rendered their role obsolete, and indeed all have now disappeared from around the Irish coast. Such sources, however, together with onshore coastal stations, have resulted in important data repositories being created at Met Éireann.

Fixed Observations

As part of a 13-station synoptic network, Met Éireann maintains five stations on, or very close to, the coast. These are manned on a 24-hour basis by trained observers who report conditions every hour. Observations include wind speed and direction, visibility, air temperature, dew-point, amount, type and height of clouds, atmospheric pressure, pressure tendency and weather conditions, as well as details of precipitation type and amount. Summaries of these observations, together with statistics on extremes are published by Met Éireann in their *Monthly Weather Bulletin*. A number of stations have also participated in programmes of chemical analysis of air and precipitation, of radioactivity levels, and of incoming solar radiation, some of which are also published in the *Monthly Weather Bulletin*. The data, which are quality-controlled, commenced mostly in the mid 1950s, and have been entered into a relational database. These coastal stations also include some of the longest records of climate available in Ireland with locations such as Valentia, Malin Head, Roche's Point and Belmullet having records spanning more than a century.

The importance of sea surface temperature (SST) for onshore weather over large areas has become much more widely appreciated in recent years, especially with the publicity accorded to recent El Niño events. Although near-real-time, satellite-derived estimates of SST for the seas surrounding Ireland can be obtained from the United States National Oceanic and Atmospheric Administration, land-based thermometer measurements are only conducted at relatively few locations. SSTs have been collected twice daily at Malin Head from 1957 to 1991 and once daily thereafter. The close correspondence between air and sea temperatures is striking (Figure 4.3) and is attributable to the North Atlantic Drift. This water takes about eight months to reach the Kerry coast from Florida, by which time its temperature in January is about 10°C, on average some 3-4°C warmer than the air over the land. Such a thermal contrast facilitates a transfer of sensible and latent heat to frontal and convective systems which are thus more active in winter along western coasts.

Five of the former fleet of eight light vessels provided wind speed (Beaufort) estimations twice daily off the eastern and southern coasts for various periods ranging from 1939 to 1982 when the last vessel, at Coningbeg, was withdrawn. Wave data were also reported from seven of the vessels over the period 1964-1975. Wind and wave data have also been recorded at the Marathon Gas Platform since 1979.

Automatic marine stations have become much more common in recent years as the data deficiencies from marine areas for input to numerical weather models have become apparent (see Chapter 8). The failure of such models to adequately predict explosive deepening of depressions is seen to be, in part, a consequence of poor input data for initialisation purposes from sea areas west of Ireland. As a consequence, a renewed interest has developed in moored buoys, automatic light vessels and other fixed platforms. The Irish Marine Data Buoy Network was inaugurated in October 2000 with the deployment of the first of a new generation of fixed buoys. The first two are located

80 km west of Inishmore and east of Dublin Bay, with further buoys intended to be deployed off the Wexford, Donegal and Kerry coasts. Reports are available on Weatherdial Fax and will become freely available on the Internet. Further afield, the UK Meteorological Office currently has 28 such stations, some as far west as 19°30' which are also used for wave model validation. Both past and near-real-time weather and wave conditions are available on the Internet from 11 of these locations around Ireland. Such buoys also have utility in other areas of coastal management nearer shore. Two buoys have been maintained since August 1998 by the Marine Institute in Bantry Bay and south of Sherkin Island, which are of particular use for forecasting blooms of toxic marine algae ("red tides").

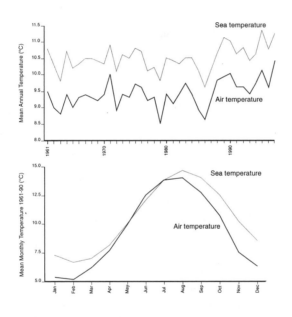

Figure 4.3: The relationship between air and sea temperature

Variable-Location Observations
A substantial data archive exists as a result of weather reports made from ships around the Irish coast. These extend back to 1854, and are more reliable after the mid 1940s. Logs were also required to be written up while ships were in port, and this often provides useful information from periods before regular meteorological observations commenced. In more recent times, ship observations usually include wind, temperature, visibility and cloud parameters, as well as SST, wave height, and swell height/period/direction. Regular voyages are particularly useful. For example, two

supply ships servicing the Marathon gas platform have provided an unbroken 4 times daily set of observations for over six years.

Among the most variable of reporting locations are a large group of drifting buoys which record air and sea temperatures, and occasionally wind. As with the moored buoys, satellite communications enable the data to be fed directly into forecast models.

Wave and Swell Data

Wave conditions are of considerable interest to activities such as aquaculture, fishing, offshore oil and gas activities, and leisure users of the coastal zone. Waves are complex phenomena, difficult to predict since they may arrive at the coast in response to a multiple set of distant events. It is now appreciated, for example, that waves breaking on the west coast may have started their journey as far away as the Caribbean. They subsequently arrive at the coast as different families, having different heights, wavelengths and speeds. Interaction between families may cancel out or reinforce the wave characteristics concerned. A computerised wave model can best disentangle this complexity, and in the case of Met Éireann, this has been done since July 1996. The WAM Model is centred on Irish waters and runs on a spatial grid of 0.25° latitude/longitude. Predictions are verified using ship and buoy data, and the ERS-2 Satellite Radar Altimeter. Nearer shore, the model accommodates itself to the bathymetry and provides a good archive of coastal wave climate. Wave energy around Ireland averages out at 50 kW m^{-2}, representing one of the highest wave energy climates in the world. Extreme value analysis reveals that the 50-year wave height off the west coast is 35 metres, over three times the equivalent value for the central part of the Irish Sea.

Swells are waves of long wavelength (300-600 m), often of only a few centimetres in amplitude which originate from disturbances perhaps several thousand kilometres distant. Swell can be a significant consideration for offshore structures where damage may occur from very low amplitude waves of a particular frequency. The WAM Model provides swell/wave energy forecasts over 25 frequencies, some corresponding to wave periods as long as 24 seconds.

Present and forecast wave conditions for the seas around Ireland are both available from the U.S. Navy. Private organisations such as Oceanweather, provide both observational data and forecasts ranging up to 7 days ahead for significant wave height and direction. The significant wave height is the average value of the vertical distance between the crest and trough of the highest one-third of all waves present.

Tides

The longest wavelength ocean waves are those associated with the twice-daily ebb and flow of the tides. After 170 years of use, the Palmer-Moray tide gauge is still the most widely used instrument for recording tidal elevation. This consists essentially of a wooden float that traces its rise and fall on a clockwork-powered rotating drum. Five tidal stations are maintained: Malin Head, Belfast, Larne, Galway and Dublin. At the discretion of the harbour-master, longer-term records may be made available. For predictive purposes, UK Admiralty Tide Tables provide specific information for Dublin which enables tidal predictions for other parts of Ireland to be published. A tidal prediction programme is also available from the UK Hydrographic Office which can provide daily tidal predictions for a large number of locations in terms of timing and heights for an indefinite period into the future.

Tidal currents are of particular concern for activities such as aquaculture, though monitoring of them is much poorer than for tidal elevation. Data are very scattered, consisting of observations of floats and poles from a variety of sources. Occasional observations are available from some of the lightships and even more rarely, some current meter data from a 10 m depth are collected. Where a tidal wave enters a restricted bathymetry, tidal current streams tend to be faster and thus eastern coasts show significantly faster tidal current streams than occur on the west coast. The British Oceanographic Data Centre at the Proudman Laboratory provides a wide range of data, as do the Admiralty Tidal Stream Atlases.

Fisheries
The oceans surrounding Ireland support a diverse range of commercial pelagic, demersal and shell fisheries. Although fish yield per unit area is only 40% that of the North Sea, the Irish fishing industry provides employment for over 15,470 people, about 60% of whom work on the fishing fleet directly, with the remainder engaged in processing and other ancillary activities. For some small rural communities, up to a quarter of the workforce may be dependent on fishing. The marine food sector is worth about 0.5% of Irish GDP.

1,385 vessels are registered to the Irish fishing fleet. The main fishing grounds are in the Irish Sea, with offshore Atlantic areas becoming important during spring and summer. Whiting, cod and seasonal species such as herring, mackerel and sprat dominate the annual catch of 300,000 tonnes valued at over 250M Euros. Most demersal fish are caught in mixed trawl fisheries with a substantial proportion of the cod and whiting catch occurring in the spring. Herring fishing in the Irish Sea is targeted on pre-spawning and spawning shoals in the autumn, especially in the vicinity of the Isle of Man. Information concerning Irish fisheries is available from the Department of the Marine and Natural Resources, or from their National Fisheries Research Centre, Abbotstown, Co. Dublin. The latter also houses specialist facilities for fish stock assessment, fish health research, and a plankton laboratory.

Landings of shellfish such as nephrops, lobster, crab and whelk have a value of 40M Euros. Nephrops is the single most valuable species in the Irish Sea and over 80% of the catches are made on the muddy seabed between the Isle of Man and the Irish coast. Mussels are dredged close to the coasts of Donegal, Wexford and Louth, though increasingly shellfish are farmed.

Access to fisheries such as the Irish Sea is strictly controlled by quota limitations under the Common Fisheries Policy. This provides for access from other EU states, subject to a complex regulatory system which includes quotas, limits on access, and other measures to prevent over-fishing and protect spawning grounds. Three distinct fishing zones are demarcated for Irish waters (the imperial unit, the mile, is the legal unit): (1) *0-6 miles* (0-9.6 km) – Republic of Ireland and Northern Ireland boats only; (2) *6-12 miles* (9.6-19.2 km) – boats from Great Britain, France, the Netherlands, Belgium and Germany are permitted to fish in particular areas for particular species; and (3) *12-200 miles* (19.2-320 km) – All EU member states and some others, including Norwegian boats.

Aquaculture currently provides employment for about 3,000 people, of whom 860 are full time employees. Some 195 fish farming operations currently exist, many in peripheral coastal regions where alternative employment is scarce. Output valued at over 125M Euros represents about 50% of Irish fish production. Eighty percent of output is export markets in France and other EU countries.

Finfish production has grown rapidly from 20 tonnes in 1980 to a peak of 14,000 tonnes in 1993. In more recent years a slight decline has set in as a result of competition from countries such as Scotland and Norway which produce around 60,000 tonnes and 300,000 tonnes respectively. Activities centre mainly on salmon production (75%) with smaller quantities of sea trout, rainbow trout and turbot.

Farmed shellfish production amounting to approximately 14,000 tonnes currently takes place around the Irish coast. Mussels account for 80% of this with oysters making up most of the remainder. More so than finfish farming, shellfish farming lends itself to part-time employment and about 85% of the workforce is part time.

Finally, the Irish seaweed industry had a turnover of approximately 5M Euros. Almost 90% of production is exported and increasingly finds its way into new niche products such as pigments and in healthcare as well as in the traditional areas of fertiliser production, and seaweed meal. The potential for improved exploitation of this marine resource is increasingly being recognised, and a major study has recently been published by the Marine Institute. It is likely that a very significant increase in production will occur in the next few years, particularly of brown seaweed (*Ascophylum nodosum*).

The presence of aquaculture is ultimately indicative of good quality coastal waters and may also have beneficial effects for local tourism. The visual intrusion of many fish farms in scenic coastal areas has however raised issues which have not yet been resolved. Equally the siting of some operations has been unsuitable because of a failure to adequately understand coastal water circulation systems, and this has led to localised pollution problems which has in turn been associated with problems for the well-being of the stocks concerned. Similarly, the introduction of migrant species from ballast water, such as the zebra mussel, and new parasites such as Bonamia, have given cause for concern, as have suspicions that the expansion of the sea lice population in the vicinity of fish farms may have been associated with the collapse of wild sea trout stocks in western Ireland.

Map Data

In the absence of an Irish hydrographic service, the main cartographic sources of information for Irish offshore areas are located in the UK Hydrographic Office, at Taunton in Somerset. This organisation started surveying in the 18th century when marine survey techniques were becoming more developed and when military concerns emerged regarding the need for accurate charts. It was not unusual during the Napoleonic Wars for eight times as many ships to be lost through running aground as through enemy actions. Since 1823, the UK Admiralty Charts have also been available to merchant shipping, and by the end of Admiral Beaufort's term as hydrographer in 1855, a wide range of marine-related publications, including tide tables, were available. Today, over 3,000 charts at a variety of scales exist and some of these encompass Irish waters.

Updating of charts in the vicinity of Ireland has focused mainly on the east coast and at a scale of 1:500,000. Some charts of the western Irish Sea have also been updated at a scale of 1:200,000. More detailed charts of the south coast, from Kinsale to Power Head, have also been produced during the past five years at scales of 1:50,000. The increasing use of marine cartography for recreational users has also led a number of commercial chart producers. The most prolific of these is probably Imray, Laurie, Norie and Wilson Ltd. who produce charts for yachtsmen and who have recently produced a number of charts for coastal areas around Ireland at scales varying between 1:150,000 and 1:280,000.

Increasingly, marine customers demand cartographic products in digital format, and the UK Hydrographic Office now supplies digital charts on CD which can be used in conjunction with a Global Positioning System. The marine areas around Ireland are included in two of the 10 CDs in this series. This method of providing map data also lends itself to regular updating by constantly providing new versions of the product incorporating the latest information such as sea depths, buoys, lights, port developments, pipelines and cables.

4.6   Soil Resources

(J.F. Collins)

4.6.1   Introduction

Soils are the interface between the geosphere and the biosphere; they are modifiers of conditions in the upper terrestrial hydrosphere and in the lower reaches of the atmosphere. Soils are the core and buffer of many terrestrial changes, resilient to a range of long-term impacts (e.g. acid deposition) but sensitive to others in the short-term (e.g. surface wetness). Soils have attributes that determine the ratio of infiltration to run-off, the ratio of absorbed to reflected energy and the circulation of gases at ground level. The importance of soils for the support of plant and animal life is unquestioned; however they may be nutrient-poor or -rich, acid or alkaline, well- or poorly-drained, but in all circumstances they are involved in a variety of cycles and feed-back mechanisms that make this world a comfortable place in which to live.

The great variation in the earth's soil mantle is closely related to the geographic patterns of climatic and biotic zones, while at national level, geology and hydrology are more influential. Terrain attributes such as gas exchange and energy partitioning are greatly influenced by topsoil properties such as colour, texture, permeability and organic matter content. Topsoil and subsoil hydrological conditions are determined by properties such as hydraulic conductivity, pan layers and root volume. Crop-ecology and production capacity are often determined by conditions in the whole profile, such as rooting volume, water holding capacity, nutrient status, drainage and depth to water table.

The soil archives and data-bases of the world are replete with information on individual points (i.e. soil profiles) of the landscape. It has been the job of soil surveyors, not just to collect such data, but to relate them to mapping units, usually soil series. It has fallen to cartographers and others to reduce these data (categorically, cartographically or some other way) into more generalised units such as soil associations or terrain (physiographic) units. In order to facilitate managers, planners and others, an increasing amount of soil data (tabular, map and air-photo) is becoming available in digital formats at various scales. What follows is a short resume of the kinds and sources of information available on Irish soils at present.

The data assembled on the soils of Ireland in the great surveys of the last century (Royal Dublin Society (RDS); Geological Survey of Ireland (GSI), Ordnance Survey (OSI), Townland Valuation), while still very interesting, are of limited scientific value. As its title suggests, Kilroe's (1907) "Soil Geology of Ireland" is a geologically-biased view of soil resources. Some reference is made to soil conditions in the Drift map series of the GSI in the earlier decades of this century (Collins, 1981). The endeavours of that institution in studying soils culminated in their report of the soils of the Department of Agriculture farm, Ballyhaise, Co Cavan, the soil map (scale 1: 5,280) of which received a prize at a Franco-British exhibition in London in 1908 (Herries Davies, 1995). An overview of the modern era, which started about 1930, is given by Cruickshank (1984), while numerous individual references are cited in Hayes's (1965) massive compilation.

4.6.2   Soil Profile Data

Data exist for many hundreds, if not thousands of Irish soil profiles. Following international conventions, typical profile descriptions usually consist of four parts: (1) *Site data*, including data on location (National Grid reference, Ordnance Survey sheet number, townland name), altitude (feet or meters), slope or gradient (in degrees or percent), aspect, surface drainage and vegetative cover;   (2) *Morphological data*, including data on profile depth, horizon thickness and arrangements; colours, structure, texture, consistence, stoniness and rooting; (3) *Compositional data*, including laboratory analysis of particle size distribution, cations, pH, organic carbon, free-iron and other parameters; and (4) *Occasional data* on trace elements or clay mineralogy. Peat soils data usually include ash content, rubbed and un-rubbed fibre, pyrophosphate index and macro-fossil data.

The horizon designation and taxonomic placement of soils change with time and with the system used (USDA, British, FAO, or other). Sources of soil profile data include: Soil bulletins of statutory bodies such as Teagasc (formerly An Foras Taluntais), Department of Agriculture Northern Ireland (DANI), Coillte Teo. and University College Dublin; theses, reports, papers of research personnel, usually associated with third level colleges; national and international journals, proceedings of conferences and workshops and similar media.

A substantial body of good information now exists for the soils of the following counties: Antrim, Armagh, Carlow, Clare, (London)Derry, Down, Fermanagh, Kildare, Laois, Leitrim, Limerick, Meath, Tipperary North Riding, Tyrone, Westmeath and Wexford. Information of similar quality is available for West Donegal, while soil maps of West Cork and West Mayo were published without soil profile data. Similar archives exist for most EU and US states. Journals which publish Irish soil data include *Irish Journal of Agricultural and Food Research; Irish Geography; Proceedings of the Royal Irish Academy; Biology and Environment* and any mainstream soil science journal to which Irish researchers submit.

Soil Sample (Test) Data

Soil testing schemes for farmers' samples have been operating since the 1940s in Ireland and since the 1930s in Northern Ireland. Under these schemes thousands of samples have been analysed yearly for properties important to growing crops. The information generally includes data on extractable nutrients such as Ca, Mg, K and P, as well as pH and lime requirement. Trace element composition, electrical conductivity and other measurements are occasionally reported for special situations such as horticultural crops.

The main sources of these data are: Teagasc (Johnstown Castle, Wexford) and Department of Agriculture, Northern Ireland (New Forge Lane, Belfast). In Ireland, commercial laboratories approved under Rural Environment Protection Scheme (REPS) are listed by Department of Agriculture, Food and Forestry (DAFF) (Anon., 1996). Summary tables and/or graphical representation of the results are occasionally published in journals, farming press, conference proceedings and similar outlets. Users of soil test data should be aware that laboratories use different extracting/analytical procedures and that the results are not always directly comparable. An example is the use of the "Olsen" method by DANI, the "Morgan" method by Teagasc and the "EUF" method by Greencore/IAS to measure available phosphorus.

Large-Scale Soil Maps (1:1,000-1:10,000 approximately)

Accurate mapping at this range of scales is limited to sites of special (pedologic) interest such as plots, fields and farms used for research and teaching (e.g. Animal Production Research Centre, Grange, Co. Meath; Agricultural and Horticultural College Farm, Piltown, Co. Kilkenny). Similar scales were used to depict the soils of parts of National Parks (e.g. Connemara, Killarney). The mapping units (series, phases, variants) are usually supported by morphological and analytical data. Specialist data such as hydraulic conductivity, micromorphology, and speciation of Fe and Al, may be reported occasionally. These soil map units do not usually carry an identifiable name but may be identified by capital letter (A, B, C...) or number (1, 2, 3....) or occasionally by both (1, 2A, 2B, C...). It should be noted that, while the O.S. topographic maps at a scale of 1:10,260, (6 inch to 1 mile) were used as field sheets by the National Soil Survey for county mapping in Ireland, soil maps of this scale were not published; the field sheets are available, however, for inspection at the Teagasc Research Centre at Johnstown Castle, Wexford, by appointment. However, as the field surveys were directed towards smaller-scale mapping, the additional unpublished information on these sheets is limited.

Medium Scale Soil Maps (1:25,000 to 1:250,000 approximately)
The most commonly available maps at this range are the 1:50,000 map series of Northern Ireland and the 1:126,000 (inch to 1 mile) maps of some of Ireland's counties. The former identify soils according to their great group (e.g. podzol) or subgroup (e.g. peaty gley) and are depicted on 17 colour sheets numbered consecutively form the NW to the SE. The soils of nine counties and parts of other counties (mapping unit: the soil series or combinations thereof) are published separately and in colour. Most of the Bulletins include, as well as a soil map, a soil suitability map and a soil drainage map. The counties published prior to 1980 are listed in the end-papers of Soil Survey Bulletin No.36, (Gardiner & Radford, 1980). Information on soil reports published since then is available from Teagasc. Soil series are named after some locality, usually where the soil was first mapped (e.g. Patrickswell) or where that soil is most extensive (Clonroche). Complex mapping units of two or three series (e.g.. Ladestown-Rathowen Complex) are common in midland counties. Phases and variants (as in Athy gravely phase; Ashbourne Shaly phase; Rathkenny sandy variant) are sometimes shown.

Ireland's county maps are each accompanied by a comprehensive bulletin, while the publication "Soils and Environment: Northern Ireland" acts as a bulletin for the soils of Northern Ireland. Northern Ireland's soil data are also available in digital format from the Ordnance Survey of Northern Ireland.

Small Scale Maps (1:250,000 and upwards)
The most widely known and used maps within this scale range are the two editions of the Soil Map of Ireland, dated 1969 and 1980 respectively, and the Peatland Map of 1981 (scale 1:575,000). The soil maps are composed partly of material generalised from the county soil maps which were completed before the dates in question, and partly from reconnaissance data for the remaining counties. The map units are mostly Associations of Great Groups and Subgroups. The extent of each is given as a percentage of the land area represented by a Principal Soil and one or more Associated Soils. Both the 2nd edition of the Soil Map (Gardiner & Radford, 1980) and the Peatland Map (Hammond, 1980) were accompanied by bulletins (now out of print). The former includes data for forty-four soil profiles; the peatland map has an elaborate legend which includes vegetational, environmental and industrial information. Simplified, generalised, monochrome sketch maps of 16 of the 17 colour maps of Northern Ireland soils are presented in Cruickshank (1997). The generalisation is based on parent material and the scale reduction is from 1:50,000 to 1:250,000.

A single-sheet map of Land Drainage Problems of the Republic, based on questionnaires returned by Department of Agriculture officers was published at the scale 1:575,000 (Galvin, 1971). Also at the lower end of this range of scales (i.e. 1:250,000) are Grazing Capacity maps of soil series of four counties (Carlow, Clare, Limerick, Wexford) published with an accompanying Bulletin by Lee & Diamond in 1972. This Bulletin also included 3 colour maps at 1:1,000,000 – a general soil map, a grazing density map, and a grazing capacity map. The county maps were based on the 1:126,000 soil series maps of each county.

Soil Maps of Very Small Scale (1:2,000,000 :5,000,000, :10,000,000...)
Maps at these scales are of educational rather than of technical value; in the case of soil they show the general outline of Soil Orders and/or Suborders. They are usually found published in atlases and textbooks with the legend substantially modified and simplified. Examples at the lower end of this range include the Soil Map of Europe at 1;1,000,000 (Commission of the European Communities, 1985) and the Soil Map of the World at 1:5,000,000) (FAO, 1975), both printed on a number of sheets. In 1991 the FAO prepared a 1:25,000,000 map of the world's soil resources as well as a generalised version at 1:100,000,000 scale. A resume of FAO soil map series is given by Meyer-Roux & Montanarella (1998). Irish examples are found in the *Atlas of Ireland* (RIA, 1979) and *Agroclimatic Atlas of Ireland* (Collins & Cummins, 1996). Maps of these scales are also used to show the national outline of the fertility status or geochemistry of soils.

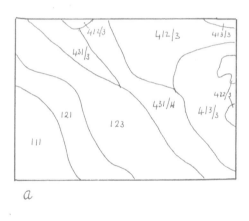

All soil maps, regardless of scale, have their uses. The scale suitable for one purpose may be entirely unsuitable for another. The ideal map of soil type for general farming may be *c.* 1:10,000; that for a research plot *c.* 1:1,000, while the county planning office would prefer scales of 1:50,000 or smaller. Figure 4.4 shows the loss of precision suffered in reducing the information recorded in the 6-inch field sheets, (a) to the scale necessary to depict the area on the General Soil Map (d). In some instances a derivative map (e.g. soil slope or soil drainage map) may be

Figure 4.4: Effect of scale on soil depiction. Scale at "a" represents linework from a 1:10:560 map. Schematically, reduction to "d" results in a scale of approximately 1:575,000

most useful; in others, soil and environmental data may be combined to create a productivity map. Regardless of scale, maps without a reliable legend and supporting field and laboratory data are of little use. Combining geographical information with numerical and descriptive data, provides a sound foundation for interpretative uses. (see Chapter 2 and section 4.8).

4.7   Vegetation Resources
(J. White)

4.7.1   Introduction

The scientific study of Irish vegetation began in 1905 with the publication of The Vegetation of the District Lying South of Dublin (Pethybridge & Praeger, 1905). Until the 1960s, however, little systematic research was undertaken, as White (1982a) has documented. The first countrywide synopsis of a vegetation type (lowland grasslands) was by O'Sullivan (1965). Since then, several comprehensive accounts of various kinds of Irish vegetation have become available, mostly as University research theses; relatively few have been published in scientific monographs or periodicals. The overall position until 1982 was summarised by White & Doyle (1982) and by other papers in White (1982b). There has been no review or synthesis of the literature since then, a period during which there has been as much research on Irish vegetation as in the preceding 80 years. In this chapter, the several relevant papers in the book edited by White (1982b) are taken as a datum point for all earlier literature, which was comprehensively surveyed therein. Sources of post-1982 information are outlined below; but given their diversity, these are merely guides rather than a complete inventory. A somewhat fuller inventory is given of maps of all periods, but this is not exhaustive.

4.7.2   Tabular Data

"Vegetation" is a collective noun for the assemblage of plant species in a particular place. Vegetation scientists often refer to this assemblage as a "plant community", on the supposition that the species interact socially in some manner, although this is a matter of considerable debate. Essential to the definition of a plant community is a listing of the species growing together in a defined space, often with a quantitative or semi-quantitative assessment of their relative abundance. Rarely is an exact numerical estimate made of populations of each species (White, 1985); more subjective estimates suffice to characterise the ensemble of species present. The listing is usually comprehensive, ideally, all vascular plants (seed plants, ferns), bryophytes (mosses, liverworts) and lichens are recorded. The sample area chosen varies with the scale of the vegetation: larger for tall, structurally complex vegetation (e.g. forests) than for lower vegetation (whether rich or poor in numbers of species). In the scientific analysis of vegetation, each particular description is regarded as a statistical sample (sometimes referred to by the French term, relevé). Repeated samples of similar vegetation are assembled into a tabular form and sorted to highlight recurrent patterns of species occurrences and coincidences. If such patterns can be detected and distinguished from other patterns in other types of vegetation, then the sorted and combined samples may be used to define a vegetation "association". This is a technical term (not synonymous with the more colloquial "plant community") to indicate a combination of species which together help to define a type of vegetation. Most significantly, not all particular, individual samples of a vegetation type may show all the defining characteristics (species) of the association; but they will have more species characteristic of that association than of any other association, defined by the same criteria.

Vegetation types are best defined not by one list of species, indeed they cannot be so defined, but by several lists combined into a tabular form (Figure 4.5). The degree of analysis of each table varies, however, with the research tradition or practical purpose of its compiler (see for example, White & Doyle, 1982). Tabular data of Irish vegetation are not necessarily precisely defined as associations; some are even poorly sorted lists of species, merely ranked from the most to the least abundant species in the data set. But the more critical investigator can use such partly sorted tables to advantage, and should be aware of them. Often, indeed, scraps of tables or even a single list of species can be combined with more comprehensive data obtained later to define or identify plant associations hitherto imperfectly known or unrecorded in Ireland (e.g. White, 1982c). There are probably about 220 vegetation associations in Ireland, of which some 150 may be regarded as being well-defined (White & Doyle, 1982). Others are provisional and need further research to confirm their status; many of these involve various types of aquatic and ruderal vegetation.

Most of the well-defined Irish vegetation associations show similarities to associations known also in Britain and north-western Europe. There is, however, some local Irish variation in species composition, typically an absence of species because of the relatively depauperate flora of Ireland (about 1350 native and naturalized species, depending on definitions. See Webb, 1978).

Vegetation scientists believe that associations reflect, in some manner, the complex interactions of climate, hydrology, soil, and biotic influences (including human management). Even when appropriate environmental data are available, it is a complex task to establish correlation between these variables and vegetation associations, except in some clear-cut instances: for example, major peatland associations are correlated with climate and hydrology, major grasslands associations with soil factors and management, some major coastal associations (of salt marshes and sand dunes) with substrate and tidal influences.

Information in tabular data
Vegetation tables are conventionally and universally arranged with species in rows and replicate samples in columns. The head of each column contains a variety of information, ideally the following: the number of the sample, aspect (compass point), inclination (slope), sample size, percentage cover of vegetation on the site, and number of species. The location of each sample is given at the foot of the table or in the text, keyed to sample number at the head of each column; additionally the date of sampling may be included. Models of this style are given by Braun-Blanquet & Tüxen (1952), which is the classic Irish exemplar and source reference (Figure 4.5). In fact, however,

few tables meet these standards, but all will minimally have the sample number at the head of each column; further information, of variable quality, on each sample may be given in the text.

An association table may list numerous (perhaps hundreds) of field samples, each one a replicate of the type of the vegetation collectively characterised by them. For convenience, this table may be condensed into a single column, indicating the sample frequency of each component species, usually on a 7-point scale. By this means, the species composition of closely related associations may be compared in a so-called "synoptic" or "constancy" table (Figure 4.6). Furthermore, such a table may be used conveniently to relate vegetation samples to those of other counties, especially in the same region, thereby highlighting geographical or ecological gradients in species composition (Figure 4.7).

Sources of data
Vegetation tables, whether formally presented as plant associations or not, are abundant for a wide variety of Irish vegetation types. More recent work of good quality usually incorporates earlier, less comprehensive tables of similar vegetation. Most of them are in unpublished Ph.D. and M.Sc. dissertations or undergraduate theses in the Universities; they are not systematically catalogued, however, and many remain unpublished. The principal sources are the Departments of Botany in University College Dublin (National University of Ireland, Dublin), National University of Ireland, Galway, and University of Dublin, Trinity College. Some have been published in whole or in part in national or international journals, principally in *Proceedings of the Royal Irish Academy B* (latterly *Biology and Environment*), *Irish Naturalists' Journal* and *Journal of Ecology*. The National Parks and Wildlife Service (now part of Dúchas, The Heritage Service) has extensive relevé data on Irish vegetation, especially on types of conservation importance. Some of these have been published (O'Connell *et al.*, 1984; Cross 1990; Crawford *et al.*, 1996); all are available to researchers. Their database allows relevés to be sorted by habitat type, vegetation type, and grid reference. During the 1970s and 1980s a variety of theses (many of them doctoral) on Irish vegetation appeared from some Dutch universities, especially the Laboratory for Geobotany, Catholic University of Nijmegen, largely inspired by Professor Victor Westhoff who maintained an active personal interest in Irish vegetation at that time. These contain extensive primary field data and local vegetation maps.

Typically, earlier datasets are incorporated into more recent theses or publications, and are often re-interpreted in the light of more comprehensive knowledge. Examples of this practice include: O'Connell *et al.* (1984; 614 relevés of wetland vegetation); Heery (1991; 267 relevés of flooded grassland (callows)); Ó Críodáin & Doyle (1994; 511 relevés of small-sedge vegetation); Kelly & Iremonger (1997; 93 relevés of Irish wet woodlands). Nonetheless, much work remains to achieve a synthesis of very diverse datasets. To date, for example, although there are now thousands of available relevés, there is no synoptic table of Irish grasslands, which account for some 65-70% of the land area. The general features, distribution and ecology of grasslands have been best summarised by O'Sullivan (1982).

Not uncommonly, accounts of British or of Western European vegetation may include some descriptions of Irish vegetation, as part of wider synopses; such sources are usually cross-referenced in relevant publications by Irish scientists. No comprehensive account of Irish vegetation yet exists.

There has long been an ideal among European vegetation scientists to develop a pan-European synthetic framework of vegetation classes. (Associations are the basic units of a hierarchical classification, successively clustered into alliances, orders and classes). Attempts to realise such a project in the past thirty years have achieved little progress, perhaps being hampered by the diversity and scale of the undertaking. Instead, there has been an increasing number of published national vegetation inventories, two of which are of particular relevance to Ireland (Rodwell, 1991-1999; Schaminée et al., 1995-1999). Clearly, a consensus on a stable description and classification of European vegetation is an essential prerequisite for trans-national vegetation mapping, nature conservation and monitoring environmental changes. This is currently being attempted by the European Vegetation Survey under the auspices of the International Association for Vegetation Science.

Small scale 1:500 000 and upwards
The *Atlas of Ireland* (RIA, 1979) contains four maps of various aspects of Irish vegetation-essentially the first multicoloured maps since 1907: (1) *The vegetation as a whole* (1:1,250,000) is indicated using major categories which account for some 90% of the non-arable cover: lowland and low-elevation grasslands, heathlands, and mires. The definition of boundaries is based on the Soils Map, compiled by the National Soil Survey, as a best approximation of these major vegetation types. This vegetation map is reproduced (c.1:2,600,000) in O'Sullivan (1982); (2) *The peatland map* (1:1M) depicts five major mire (bog and fen) types, together with the extent of their exploitation. This map (1:575,000) was revised subsequently by the National Soil Survey (Hammond, 1979); an amended version has been reproduced (1:2,000,000) by Aalen et al. (1997). A revised map of raised bogs (1:575,000) which accompanies a report by Cross (1990) carefully documents their serious depletion in recent decades; it is based on data collected during 1982-1987; (3) *A field boundaries map* (1:1,250,000), excluding unenclosed and upland areas, indicates the species composition of hedges, mostly planted since the early 18th century, although some may represent the linear fragments of long-felled woodlands. The distribution patterns, based on extensive local sampling, are derived from the General Soil Map of Ireland; and (4) *A State forests map* (1:1,250,000) depicts essentially conifer plantations in public ownership in 1972, but this is seriously outdated.

4.7.3   Vegetation Maps

Large-scale (1:1,000-1:10, 000 approximately)
Accurate mapping at these scales is typical of sites of special botanical interest. Such maps commonly occur in unpublished research dissertations in University Departments. They exist also for grassland surveys conducted by Teagasc (formerly An Foras Talúntais) on various farms (e.g. Jaritz & Lee, 1968; O'Sullivan, 1986, both at 1:2,500), and for some sites of scientific interest (Dúchas, National Parks and Wildlife Service).

	Nr. der Aufnahme	58	70	73	74	142	149	150	246	299	313	321
	Autor	Tx	BB	Tx	Tx	Tx	BB	BB	Tx	Tx	Tx	Tx
	Mähwiese (M), Weide (W)	W	MW	W	W	MW	—	W	M	M	W	W
	Exposition	N	—	SW	.	.	.	S	N	SE	SE	S
	Neigung	5⁰	—	5⁰	.	.	.	2⁰	5⁰	20⁰	3⁰	3⁰
	Grösse der Probefläche (m²)	30	100	50	.	.	.	100	.	.	.	.
	Artenzahl	33	22	30	25	29	15	20	34	38	23	26
Charakterarten:												
Hc	Cynosurus cristatus L.	1.2	3.2	2.2	2.2	2.2	2.1	3.2	2.2	2.2	.	3.2
Hrept	Trifolium repens L.	3.3	2.2	3.3	2.2	1.2	2.2	2.2	2.2	1.2	2.2	3.3
Hs	Senecio Jacobaea L.	+	+	2.1	+.2	.	.	.	.	.	+.1	+.1
Hc	Phleum pratense L.	.	.	.	.	.	.	.	.	.	.	.
Differentialarten der Assoziation:												
Hs	Centaurea nigra L.	.	2.1	+.1	+.1	+.2	.	.	+	1.2	.	1.1
Hros	Leontodon taraxacoides (Vill.) Mér.	.	.	.	.	.	.	2.1	.	.	1.1	.
Differentialarten der Untereinheiten:												
Chr	Cerastium caespitosum Gilib.	1.2	+	1.2	1.1	1.1	1.2	.	+.1	1.1	+.2	2.1
Hc	Lolium perenne L.	1.2	+	1.2	+.2	2.2	3.2	.	.	.	+.2	2.2
Hc	Dactylis glomerata L.	.	+	.	.	.	.	1.2	.	.	.	+.1
T	Crepis capillaris (L.) Wallr.	+.1	.	.	1.1	2.1	1.1	2.1	+.1	1.1	+	.
Brr	Brachythecium rutabulum (L.) Br. eur.	2.2	.	1.2	.	1.2	.	.	2.2	2.2	2.2?	2.2
Ordnungscharakterarten:												
Hs	Chrysanthemum leucanthemum L.	.	2.1	(+.1)	2.1	2.1	.	+.1	+.1	2.2	+	.
Hros	Bellis perennis L.	1.1	.	1.1	.	1.1	2.2	2.1	.	.	.	+
T	Trifolium dubium Sibth.	2.2	.	+.2	.	2.2	.	+	1.1	+.2	.	.
T	Bromus mollis L.	.	.	.	.	1.1	.	.	1.2	+.1	.	.
Hc	Trisetum flavescens (L.) P.B.	.	.	.	.	.	1.1	2.1	.	.	.	+.2
Hs	Daucus carota L.	.	.	.	.	.	.	.	.	.	.	.
Klassencharakterarten:												
Hs	Prunella vulgaris L.	2.2	2.1	2.2	2.1	1.1	.	+	+.1	2.2	+.1	2.1
Hc	Holcus lanatus L.	2.2	1.1	2.2	2.2	1.2	1.1	2.1	2.2	2.2	2.2	+.2
Hs	Trifolium pratense L.	.	2.2	+.2	1.2	2.2	2.2	+	1.3	2.2	.	2.2
Hs	Rumex acetosa L.	1.1	+	.	+.1	1.1	1.1	.	.	1.1	.	+.1
Hc	Festuca rubra L. ssp. eu-rubra Hackel var. genuina Hack.	1.2	+	2.2	1.2	1.2	.	.	.	1.2	.	.
Hs	Ranunculus acer L.	.	.	(+)	+.1	1.2	1.1	.	.	.	.	+.1
T	Rhinanthus minor L. s.str.	.	.	.	.	1.1	.	+	2.1	+	.	.
Hc	Poa trivialis L.	+.1	.	.	.	+.1	.	.	1.2	1.1	.	1.1
Hsc	Vicia cracca L.	.	.	.	.	.	.	.	+.2	.	.	.
Hc	Juncus effusus L.	2.2	.	.	.	.	.	.	+	.	.	.
Hs	Cirsium palustre (L.) Scop.	+	.	.	.	.	.	.	.	.	.	.
T	Rhinanthus glaber Lam. s.str.	.	.	.	.	+.2	.	.	.	2.1	.	.
Hs	Lotus uliginosus Schkuhr	.	.	.	.	.	.	.	.	1.2	.	1.2
Begleiter:												
Hros	Plantago lanceolata L.	1.1	2.1	+.1	1.1	2.2	1.1	+	2.2	1.1	2.2	2.1
Hc	Agrostis tenuis Sibth.	2.2	+	3.2	2.2	.	.	.	2.2	2.2	3.2	2.2
Hc	Anthoxanthum odoratum L.	1.2	+	.	.	1.2	.	.	2.2	2.3	2.2	+.2
Hros	Hypochoeris radicata L.	.	2.1	2.1	.	.	+	.	1.2	1.1	1.1	+.1
Hs	Lotus corniculatus L.	2.2	.	1.2	1.2	.	.	.	.	.	1.2	+.2
Hrept	Ranunculus repens L.	1.2	.	2.1	.	+.1	.	.	+.1	1.1	+.1	2.2
Hs	Achillea millefolium L.	.	.	.	.	.	.	.	+	.	1.1	+.1
Hros	Leontodon autumnalis L.	+	.	.	+.1	+.1	1.1	.	.	+	.	.
Chp	Rhytidiadelphus squarrosus (L.) Warnst.	.	.	2.2	.	.	.	.	+	2.2	3.3	.
Hs	Ranunculus bulbosus L.	.	+	.	+.1	.	.	+	.	.	.	.

Figure 4.5: Facsimile extract from Table 29 of Braun-Blanquet & Tüxen (1952) showing the classical layout of a sorted vegetation table. Each column represents a single field sample. Each species is given a 2 digit cover or abundance code and a morphological character code. Appropriate theory and methodology can be found in Mueller-Dombois and Ellenberg (1974)

	Puccinellietum maritimae	Halimionietum portulacoidis	Juncetum gerardii	Junco maritimi-Oenanthetum lachenalii	Artemisietum maritimae	Blysmetum rufi	Puccinellietum distantis	Atriplici-Agropyretum pungentis	Halo-Scirpetum maritimi
Number of relevés	225	22	159	13	8	11	2	3	17
Association character species									
Puccinellia maritima	V	IV	II		II	+	III		I
Halimione portulacoides	I	V	R					IV	
Juncus gerardii	R		V	II	II	V			III
Juncus maritimus	R		II	V	I	+			
Oenanthe lachenalii			R	V					
Artemisia maritima	R				V				
Blysmus rufus			R			V			
Puccinellia distans							V		
Spergularia marina	R		R	I			V		
Elymus pycnanthus								V	
Scirpus maritimus	R						III		V
Sagina maritima									
Asteretea tripolii									
Class character species									
Aster tripolium	IV	III	III	I	III		III	IV	III
Plantago maritima	IV	II	IV	III	IV	V		II	I
Triglochin maritima	II	II	III	I	I	III	III		III
Order species									
Armeria maritima	IV	I	IV		II	V			+
Glaux maritima	III	+	V	V	II	V	III		II
Alliance species									
- Puccinellion maritimae									
Spergularia media	IV	III	I						
- Armerion maritimae									
Festuca rubra	R	R	IV	V	V	IV		IV	I
Agrostis stolonifera	R		III	V	III	V	III		III

Figure 4.6: Edited extract from Table 28 of Wymer (1984) showing the layout of a synoptic or constancy table of Irish salt marsh vegetation. Each column is a vegetation association and each species is given a "constancy value" representing frequency. Reproduced with the permission of E. Wymer.

CLASS	OXYCOCCO - SPHAGNETEA BR.-BL. ET TX. 1943																											
ORDER	ERICETALIA TETRALICIS				SPHAGNETALIA MAGELLANICI																							
ALLIANCE	ERICION				ERICO-SPHAGNION										SPHAGNION FUSCI													
COLUMN NUMBER	1	2	3	4	5	6	7	8	9	10	11	12	13	14	15	16	17	18	19	20	21	22	23	24	25	26	27	28
NO. OF RELEVES	230	4	7	42	12	28	28	91	36	90	80	106	138	213	20	12	478	10	375	94	36	18	98	271	54	280	109	175
CLASS CHARACTER SPECIES																												
Drosera rotundifolia	III	III	I	III	V	IV	IV	IV	V	V	III	IV	IV	II	II	V	IV	III	II	II	.	II	IV	V	I	II	+	III
Aulacomnium palustre	+	.	II	II	V	I	.	R	.	II	I	III	III	I	.	.	R	.	II	II	II	I	.	I	I	III	IV	R
Sphagnum tenellum	I	IV	V	I	I	I	IV	IV	V	IV	II	II	+	II	II	X	II	.	I	+	I	.	I	I	.	.	.	III
Calypogeia trichomanis	R	R	.	I	.	.	I	.	.	I	+	.	.	+	.	.	.	IV	.	.	I	.	R	.	R	.	.	.
Lepidozia setacea	R	III	I	I	I	II	II	II	II	III	I	I	+	I	.	X	I	II	+	R	+	.	IV	I	.	.	.	I
Sphagnum capillaceum	+	.	I	II	.	.	.	II	.	III	I	II	I	II	II	X	II	III	I	I	.	II	I	I	II	II	I	.
SPHAGNETALIA CHARACTER SPECIES																												
Eriophorum vaginatum	II	.	.	I	IV	II	V	III	V	V	IV	V	III	V	III	V	IV	V	V	V	V	V	V	V	IV	V	III	V
Sphagnum magellanicum	R	.	.	.	III	III	I	II	III	III	IV	V	IV	R	II	X	III	V	III	IV	+	V	III	III	R	IV	IV	III
Sphagnum rubellum	R	.	.	I	V	III	IV	IV	V	V	II	III	II	IV	I	V	III	V	IV	II	III	I	IV	I	.	I	.	III
Mylia anomola	I	.	.	.	III	I	II	+	R	II	II	IV	+	+	I	X	II	.	II	I	II	.	V	IV	IV	II	R	+
Oxycoccus quadripetalus	R	.	.	.	IV	IV	.	.	.	III	V	V	V	I	.	V	V	V	IV	V	.	V	V	V	.	IV	III	V
Polytrichum strictum	R	.	.	.	V	II	.	.	.	I	I	I	II	+	I	.	R	III	II	II	.	IV	I	III	IV	IV	II	I
Andromeda polifolia	R	.	.	.	.	I	.	.	+	V	IV	V	IV	I	I	V	V	II	IV	IV	.	III	V	V	IV	V	I	V
Sphagnum recurvum	.	R	.	II	III	V	.	+	.	.	III	IV	IV	.	I	.	.	III	II	III	R	V	II	V	I	V	V	II
Pohlia nutans	+	.	.	.	.	.	.	.	.	.	+	III	+	+	.	X	+	.	+	I	+	.	+	I	+	I	I	R
Carex pauciflora	.	.	.	R	I	I	I	.	.	.	.	.	.	.	.	.	R	.	+	I	.	I	+	I	+	II	R	IV
Cephalozia connivens	.	.	.	+	.	I	.	+	.	.	R	R	R	.	.	.	.	I	R	R	.	I	I	I	.	.	.	.
Cephalozia macrostachya	+	.	.	.	.	.	I	.	.	.	+	+	+	.	.	.	.	.	R	.	.	.	+	R	.	.	.	.
Calypogeia sphagnicola	.	.	.	R	.	.	.	.	.	.	.	R	.	.	.	I	R	.	I	.	.	.	+	.	.	R	.	III
ERICETALIA CHARACTER SPECIES																												
Erica tetralix	V	V	V	V	I	V	V	V	V	V	V	V	V	IV	II	X	II	.	.	.	I	.	.	.	.	.	.	.
Trichophorum aespitosum	III	+	IV	IV	.	II	V	III	V	IV	I	.	R	IV	V	.	.	I	IV	.	II	.	.	.	.	.	.	.
Sphagnum compactum	III	V	IV	+	.	I	.	.	.	.	+	II	R	.	IV	.	+	.	.	.	.	.	.	.	.	R	R	II
Juncus squarrosus	III	.	III	II	III	I	II	.	.	.	.	.	.	I	III	.	.	.	+	R	III	.	.	.	.	.	.	.
ERICO-SPHAGNION CHARACTER AND DIFFERENTIAL(*) SPECIES																												
Sphagnum papillosum	+	+	I	IV	III	V	III	IV	V	V	III	+	II	II	I	I	I	.	R	R	I	.	.	.	.	R	.	V
Odontoschisma sphagni	I	I	II	I	III	II	IV	III	V	IV	III	II	II	II	.	I	II	+	I	.	R	.	+	.	.	.	.	.
*Molinia coerulea	IV	V	V	IV	V	V	IV	V	II	.	II	I	III	R	II	.	.	.	.	+	.	R	.	.	.	.	.	.
*Eriophorum angustifolium	III	IV	I	IV	V	IV	V	V	V	V	IV	V	V	V	.	.	I	.	+	R	II	.	.	.	+	.	R	R
*Hypnum cupressiforme	III	II	IV	IV	.	III	.	II	II	IV	II	II	I	II	.	.	.	.	.	.	II	.	.	.	.	.	.	.
*Cladonia impexa	II	IV	V	II	I	I	IV	IV	IV	V	II	II	R	III	.	.	.	.	.	.	I	.	.	.	.	.	.	.
Narthecium ossifragum	I	IV	II	IV	.	V	V	V	V	V	III	.	R	II	.	I	II	.	.	.	.	.	.	.	.	.	.	.
Sphagnum plumulosum	R	.	.	II	.	I	IV	II	II	II	R	I	R	II	.	I	.	.	.	.	II	.	.	.	.	.	R	.
*Campylopus flexuosus	R	I	I	I	.	.	.	III	.	II	.	.	.	II	.	.	.	.	.	.	I	.	.	.	.	.	.	.
*Rhynchospora alba	I	IV	.	+	.	.	.	III	II	III	II	R	I	.	.	.	II	.	.	+	.	.	+	+	.	.	.	I
Myrica gale	R	+	.	I	.	.	III	II	I	R	II	.	.	.	.	.	.	.	.	.	.	.	.	.	.	.	.	.
*Leucobryum glaucum	II	.	III	+	.	.	.	I	.	+	I	.	R	R	III	.	.	+	.	R	.	.	.	.	.	.	.	.
Sphagnum imbricatum	.	.	.	.	.	I	.	I	I	II	.	.	.	.	.	R	R	.	.	.	.	.	.	.	.	.	.	.
Drosera intermedia	R	I	.	.	.	.	.	I	.	R	R	.	+	.	.	.	+	.	.	.	.	.	.	.	.	.	.	.
SPHAGNION FUSCI CHARACTER AND DIFFERENTIAL(*) SPECIES																												
Sphagnum fuscum	.	.	.	.	.	.	.	R	R	II	.	.	.	R	I	.	II	I	I	I	II	II	V	V	V	IV	I	+
*Empetrum nigrum	.	.	.	.	.	.	.	.	I	+	I	I	+	III	.	X	I	.	III	II	II	.	IV	V	I	III	II	I
Rubus chamaemorus	.	.	.	.	.	.	.	.	.	.	.	.	.	R	.	V	II	.	.	.	V	.	V	V	V	IV	IV	+
*Cladonia rangiferina	.	.	.	+	.	.	.	.	.	.	.	.	.	R	I	.	II	IV	I	I	II	.	V	IV	IV	III	I	+
*Cladonia sylvatica	.	.	.	.	.	.	.	.	.	.	.	.	.	.	I	X	I	.	I	I	IV	.	I	III	IV	III	I	.
*Vaccinium uliginosum	.	.	.	.	.	I	.	.	.	.	.	.	.	.	II	.	R	III	II	III	II	II	II	IV	IV	V	III	I
*Vaccinium vitis-idaea	.	.	.	.	II	I	.	.	.	.	.	.	.	R	I	.	.	I	I	II	IV	R	.	+	IV	IV	V	.
*Betula nana	.	.	.	+	.	.	.	.	.	.	.	.	.	.	.	.	.	.	R	.	II	.	I	III	V	IV	R	III
Ledum palustre	.	.	.	.	.	.	.	.	.	.	.	.	.	.	.	.	.	.	R	.	.	V	+	III	IV	V	IV	R
Trichophorum austriacum	.	.	.	.	.	.	.	.	.	.	.	.	.	.	.	.	III	.	.	R	.	.	III	R	R	R	R	V
Oxycoccus microcarpus	.	.	.	.	II	.	.	.	.	.	.	.	.	.	.	.	.	.	.	.	I	+	IV	IV	IV	III	II	I
Empetrum hermaphroditum	.	.	.	.	.	.	R	.	.	.	.	.	.	.	.	.	.	.	.	.	V	.	.	.	V	III	II	+
Dicranum bergeri	.	.	.	.	.	.	.	.	.	.	.	.	.	.	.	.	+	.	I	I	.	.	II	III	II	I	+	R
Chamaedaphne calyculata	.	.	.	.	.	.	.	.	.	.	.	.	.	.	.	.	.	.	.	.	.	.	.	I	.	II	I	+
Cephalozia media	.	.	.	.	.	.	.	.	.	.	.	.	.	.	.	X	I	II	.	.	.	.	II	R	+	.	.	.
Cetraria islandica	+	.	.	R	.	.	.	.	.	.	.	.	.	.	I	.	+	.	I	.	II	.	I	R	I	R	R	R
Cladonia squamata	.	.	.	R	.	.	.	.	.	.	.	.	.	.	I	.	.	.	+	.	I	.	I	+	R	.	R	R
Cladonia alpestris	.	.	.	.	.	.	.	.	.	.	.	.	.	.	.	.	I	.	.	.	.	.	I	II	III	I	R	.
Calypogeia neesiana	.	.	.	.	.	.	.	.	.	.	.	.	.	.	.	.	.	.	R	R	.	.	I	+	+	.	.	.

Figure 4.7: Edited extract of Table 1 from Moore (1968) showing the layout of a synoptic or constancy table of bogs and wet heaths of northern Europe. The table summarises data on 3135 relevés throughout the region. Columns 4, 8, 10 and 14 include data from Ireland

Published maps at this scale include, for example, vegetation of sand dunes and salt marshes at both North Bull Island (c.1:8,750) (Moore & O'Reilly, 1977) and Malahide Island, Co. Dublin (c.1:10,500) (Ní Lamhna, 1982); woodland vegetation at Derryclare, Co. Galway (c.1:8,000) (Ferguson & Westhoff, 1987); aquatic and wetland vegetation of Lower Lough Corrib, Co. Galway (c.1:8,000) (Mooney & O'Connell, 1990), and a polychrome map of vegetation around Mullach Mór, Co. Clare (1:7,500) (Moles & Travers, 1983).

Medium-scale (1:20,000-1:250,000 approximately)
The two earliest maps of Irish vegetation were at a medium scale. A map of an area (some 200 square miles) lying south of Dublin depicted 13 vegetation types at a scale of 1:63,360 (Pethybridge & Praeger, 1905); the original field-maps (1:10,560) are still available (National Botanic Gardens Library). The map was widely acclaimed for its technical excellence and was the first vegetation map printed (at the Ordnance Survey Office, Southampton [UK]) and published by the British government (White 1982a) Together with a map of vegetation (1:21,120) on Lambay Island, Co. Dublin (Praeger, 1907), these maps constitute the only polychrome vegetation maps published in Ireland until the 1970s. A map of the vegetation of Clare Island, Co. Mayo (1:31,680) (Praeger, 1911), depicting eight vegetation types, concluded the series of vegetation maps published by Praeger. Some further sources of maps made during the 1920s and 1930s by botanists at The Queen's University, Belfast are reviewed by White (1982a). A map of the vegetation of Carrowkeel, Co. Sligo (1:26,000) (Webb, 1947) is an isolated example during a long fallow period of Irish vegetation mapping, not revived until the 1960s. Moore (1960) resurveyed and remapped an upland area of 100 km^2, part of the district originally mapped by Pethybridge & Praeger (1905); his map (1:38,500) documented changes that had taken place in the intervening 50 years, and is the only published example of such a comparison based on mapping.

Peatlands in Northern Ireland have been comprehensively mapped at 1:20,000 using air photographs (Cruickshank & Tomlinson, 1990). Generalised vegetation categories have been recorded in the broad survey, but some field-based validation of the exact botanical composition of the air-photo images has been conducted, resulting in a vegetation map (1:26,500) of part of the Garron Plateau, Co. Antrim (Tomlinson, 1984).

It may be noted that localised maps of peatlands (at various scales) were published in the Reports of the Commissioners appointed to enquire into the nature and extent of several bogs in Ireland (1810-1814); but they treat peatlands as an economic resource, with little reference to their botanical variety. Examples have been reproduced in historical accounts of Irish peatlands (e.g. Feehan & O'Donovan, 1996; Aalen et al., 1997). An attempt to represent the forests of Ireland (c. 1:1,860,000) (conifer plantations, broadleaf, and mixed) is given by Aalen et al. (1997). This is derived from the CORINE database of satellite imagery (with a minimum identifiable unit of 25 ha), and is the most comprehensive forest map now available at this scale. Woodlands in Co. Wicklow are represented (c. 1:590,000) in Aalen et al. (1997).

The CORINE database, used for the forest map mentioned above, employed a gross level habitat classification scheme based largely on plant formations well defined in western Europe (Devillers *et al.*, 1991). This permits accurate mapping (1:100,000) of major vegetation types, subject to field verification, but lacks the level of discrimination normally demanded by a vegetation scientist. Maps have been published at 1:500,000, and a composite (1:2.5M) is reproduced by Collins & Cummins (1996). Maps for specified areas are available to order commercially from ERA-Maptec, Dublin, which holds the CORINE database. Pasture grasslands, forests, heathlands, and peatlands, all of varying botanical composition, are discernible. It is not simple to relate the CORINE categories to particular vegetation associations; some of the problems involved have been outlined by Cruickshank & Tomlinson (1996). The Forest Service has developed a Forest Inventory and Planning System (FIPS) using a comprehensive combination of data bases, with satellite imagery as a primary source: all forest lots greater than 0.2 ha have been classified in one of twenty categories. Information may be obtained from FIPS Management Unit, Forest Service, Johnstown Castle, Wexford.

Ireland is represented on the vegetation map of the Council of Europe member states (1:3M) (Ozenda, 1979), but this map is grossly misleading, based as it is on a metaphysical conceit: the natural vegetation of Ireland as it might be in the future absence of man! Most of the island is shown covered by forests (ash or oak predominantly), heathlands and bogs, but even the latter (which are common) are poorly and inaccurately circumscribed. Further, more reliable refinements of this map have been made (Cross, 1998) but on the same principles; consequently, whatever its theoretical virtue, the map has dubious practical utility.

Distribution Maps
Inventories of sites have been occasionally compiled for particular vegetation types which are relatively infrequent or of small extent when compared to the predominant grassland cover of Ireland. Sometimes these are shown on dot distribution maps: for example, wet woodlands (Kelly & Iremonger, 1997), woodlands on esker ridges (Cross, 1992), wetland vegetation (O'Connell *et al.*, 1984). The distributions of most of the sand dunes (Curtis, 1991) and of salt marshes (Curtis & Sheehy Skeffington, 1998) have been mapped, thereby indicating the localities of potential vegetation types which are commonly found in these habitats. A comprehensive inventory of raised bogs is provided by Cross (1990). Dúchas, The Heritage Service has unpublished inventories of blanket bogs, turloughs and esker grasslands.

Some information on the distribution of vegetation types may be gleaned from *Atlas of the British Flora* (Perring & Walters, 1962): this shows for each species in the Irish flora its distribution within a 10 km x 10 km grid (c. 900 of which cover Ireland). The restriction of a particular species to a particular vegetation may, with the help of the distribution map, define the occurrence of the vegetation, at least at a relatively crude scale. But these are at best only guidelines, given the vagaries of species' distributions. Moreover, the recorded distributions may be imperfect; a revised edition of the Atlas is projected for publication in 2001, containing updated and expanded records.

4.8 Land information and appraisal of land resources

(J. F. Collins)

4.8.1 Introduction

Acquiring basic information on rocks, water, vegetation, soils and other resources is not an end in itself. The information is much more useful when individual parts are combined, analysed and interpreted to get "added value". Towards the end of the 20th Century there were major advances in land evaluation (e.g. concepts such as resilience and sustainability) which have benefited our understanding and management of land resources. To maximise these new-found benefits, particular attention must be given to efficient methods of archival and retrieval of data, as well as to the computing capacity for its processing and modelling. Beek (1978) purported that the concept of land quality emanated from the desire to reduce the enormous amount of land data to manageable proportions without loss of information. In its revised and enlarged *Soil Survey Manual*, the Soil Survey Division Staff of the US Department of Agriculture (Soil Survey Staff, 1993) devoted 170 out of 450 pages to the task of bringing soil survey information to the user. Chapter 6 (146 pages) deals with "Interpretations", while Chapter 7 is entitled "Disseminating Soil Survey Information". In the same year , 1993, the Land and Water Development Division of the FAO published a 250 page book in response to increasing demand for adequate information on computerised systems for land resources appraisal for sustainable agriculture (Chidley *et al.*,1993). In their introduction the authors drew attention to the issue of climate change and the need for timely, reliable and meaningful information on land resources potentials and limitations. More recently the European Soil Bureau (Heineke *et al.* 1998) published a 550 page treatise on land information systems, devoting sections to both national and European perspectives. In the latter section the authors discuss information policy, access to European databases, environmental applications and land evaluation.

Starting with the publication of its first Bulletin in 1964, Soil Survey Bulletins of the National Soil Survey of Ireland include a chapter headed "Soil Suitability". The preamble usually emphasises that the ratings assigned were qualitative rather than quantitative. In the 1970s and 1980s the ratings became more quantitative on foot of experimental data and yield measurements of grass, arable, forest and other crops. In their digest of Irish soil resources, Gardiner & Radford (1980) stated that Chapter 11 (pages 125-141) "attempts to interpret for practical use the basic data derived from the Soil Map." As well as grouping 44 Soil Associations according to physiographic location and administrative area, Irish land was classified either as "Marginal" or "Tillage", the latter being divided into four subclasses: highly suitable; suitable; moderately suitable; and marginally suitable.

4.8.2 Land resource appraisal

Chidley *et al.* (1993) define land as "An area of the earth's surface, the characteristics of which embrace all reasonably stable, or predictably cyclic, attributes of the biosphere vertically above and below this area including those of the atmosphere, the soil and underlying geology, the hydrology, the plant and animal populations, and the results of past and present human activity, to the extent that these attributes exert a significant influence on present and future uses of the land by man". They listed 13 applications of land resources appraisal or evaluation, stating that the outputs are used by planners, economists, engineers, scientists, politicians and others. The most significant applications require quantitative knowledge of the input data – the kind of data that are the main focus of this chapter. These applications are listed below with a note attached to each showing current or past examples in Irish contexts where such data were used. Examples help to show how data from a wide variety of sources, collected by personnel from disparate institutions and for widely differing purposes can be assembled, reprocessed and interpreted for an ever-widening range of uses.

Land Suitability Assessment
Despite the early attempt by Storie (1933) to express soil/land quality numerically, ranking on an ordinal scale (e.g. low-medium-high) remains commonplace. Most rankings are connected to the use and management of land for production purposes, consistent with environmental conservation. Using climate, landform, crop yield and other data, most Irish county soil survey reports include tables and maps of suitability. The earlier of these indicated suitability for "all purposes"; more recent ones give separate ratings for general tillage, grassland and forest crops. Usually 5 or 6 Suitability Classes are recognised and their distribution shown on an accompanying map by categoric generalisation of the soil map units. The national picture was portrayed in map form by the Royal Irish Academy (RIA, 1979), and outlined in tabular and text form by both Gardiner & Radford (1980) and Cruickshank (1997). There is as reasonable degree of similarity between "Tillage Classes 1, 2, 3 & 4" of Gardiner & Radford and "Agricultural Land Grades 1, 2, 3A & 3B" of Cruickshank, and between "Marginal Land" of the former authors, and "Grades 4 & 5" of the latter. Both classifications are based on modifications of the original guidelines of Klingebiel & Montgomery (1961). The geographic distribution of suitability classes for tillage and grazing were depicted on maps of very small scale in recent AGMET publications (Keane, 1986; Collins & Cummins, 1996).

Land Productivity Assessment
Maps and tables compiled under this heading are generally quantitative; the data are given in units such as tons of dry matter/ha or livestock units/ha. Examples include: 6 classes of potential forestry yield in cubic metres per hectare per annum, and grazing capacity/stocking rates in Livestock Units per hectare under low and high nitrogen regimes. Irish data were summarised for the soils of four counties (Lee & Diamond, 1972) and continued in subsequent soil survey reports. Lee (1986) placed Irish grassland in 3 productivity classes (10,000-12,000; 6,000-12,000, <6,000 kg dry matter /ha) on the basis of moisture availability, poaching susceptibility, and access for machinery. When included, the scale of productivity maps may vary form 1:2,500 (institutional farms) to 1: several million (countrywide) i.e. they are governed by the scale of the main input which is usually a soil map. Research supporting these kinds of assessments includes that of Brereton (1972), Brereton & Keane (1982), Conry (1985,1996), Conry & Hegarty (1997), Lee & Ryan (1966) and McEntee (1979).

Population Support Capacity
While this heading is meant to assess land availability for human habitation, the concept could be extended to make assessments such as grass growing season (Connaughton, 1973; Betts, 1982), grazing season (Keane, 1998), risk of water pollution from septic tank disposal systems (Daly *et al.*,1993), and to the number of machine work days per year. The concept may also include the effect of cattle, pigs and poultry, fish farms, surface and groundwater supply, and even intensive conifer planting (Cullinan & Bulfin, 1996), on the environment. Such interpretations are becoming common-place in Environmental Impact Statements and are prerequisites for awarding Integrated Pollution Control licences in many instances (EPA, 1995a, b).

Land Evaluation and Land Use Planning
While there are virtually no limits to this theme (Convery & Feehan, 1995), major areas of interest under this heading in Ireland include: designation of Special Areas of Conservation (SAC) and Natural Heritage Areas (NHA); location of National Parks, zoning for development, (use of cut-over peatlands, location of sanitary landfills, golf courses, and arterial drainage schemes). All of these require an assessment of climatic, soil, hydrological, socio-economic, amenity and aesthetic parameters. The purchase of turbaries and turbary rights with a view to conserving even very small parcels of peatland needs accurate ground survey and large scale maps. However, the scientific bases which underpin some designations or delineations (especially the SACs) are often unspecified and lead to public disquiet. Using a much smaller scale view, Gardiner discussed the value of soil survey in regional development (1981) and climate modelling (1982).

Land Degradation Risk Areas
In common with many terms dealing with the environment, the word "degradation" has many meanings (McIsaac & Brun 1999). When used with "land" it involves defining soil qualities, attributes and processes, and introduces concepts of resilience and sustainability (Taylor et al., 1996). In Ireland examples of research effort (and concern) include atmospheric deposition and accelerated acidification (acid rain, conifer litter) (Farrell,1995), peat instability on slopes, excessive grazing, soil surface crusting (Hussain et al., 1985), subsurface compaction (Larney 1985), and water quality (Sweeney, 1997). Critical factors in understanding the causes and processes involve detailed knowledge of climate, soil and landforms (Wilcock, 1997).

Quantification of Land Resource Constraints
Constraints of a heritage or amenity nature are difficult to quantify; others are much easier but the variables may be of a long-term nature (return periods of droughts, floods, severe frost). Depending on location, constraints may include: nutrient deficiencies/plant health (Stanley *et al.*, 1996), declining carbon pools (Cruickshank *et al*, 1998), aquifer vulnerability (Daly & Warren, 1998), runoff risk (Sherwood, 1992) availability of irrigation water, grounds for waste disposal (Daly,1998), exposure (fish farming, tree-throw), migratory bird corridors and sanctuaries, amenity/aesthetic concerns (wind turbines, communication masts) and disease carriers (a wide range of biotic species). A combination of some of these constraints was used in compiling site types for afforestation (Cummins & Whelan, 1996). A comprehensive Irish study on

resource constraints (i.e., atmospheric inputs/forest health) was reported on by Boyle *et al*. (1997). However, not alone are we still lacking quantifiable data on most of the constraints listed above, but the guidelines necessary to measure them are also critically wanting.

Land Management
The everyday implementation of a land use (or farm) plan may need readily available information (on computer) and feed back, especially if a change has to be made due, for example, to unpredicted weather conditions. Day-to-day, even hourly, decisions must be made in light of weather outlooks as to timing of fertilising, sowing, spraying, mowing and many other weather-related farm operations. Large intensive enterprises such as pig and poultry units must keep daily records of where slurry wastes are being applied, and have contingency plans for possible outbreaks of certain diseases. Farmers in Rural Environmental Protection Schemes (REPS) who must farm within restrictive guidelines are among those in need of information and feedback (Maloney, 1994). The concept could be enlarged to include "crisis management" in events such as oil spills, fish kills, fire damage and related accidents.

Agro-ecological Characterisation for Research Planning
Since the findings of field research should have the widest possible applicability, the location of research stations, farms, catchment basins, monitoring sites and even single sampling sites must be chosen with reference to the most up-to-date information on the major components - water, soil, energy, ecology. It is economically ineffective to invest a large research effort in a site which extends to, say, 1% of a region when equally applicable results could be acquired in an adjacent site which represents 5%, 10% or more of the region in question. In acquiring the information needed to make the correct choice of site, the primary source/form of the data should be searched for, since a lot of information is lost in transferring to a smaller scale (through cartographic or categoric generalisation). For example, the 2-sheet CORINE map of Ireland (1:500,000) shows a national land-cover picture but the local details should be assessed by reference to the original data (O'Sullivan *et al.*, 1994). Thematic maps which are built on detailed base maps facilitate locational accuracy and are superior instruments in interpretation and planning routines.

Technology Transfer
Three examples are given:
(a)   Agricultural advisors/consultants should have access to the data needed to create a land resource inventory of their area. A combination of the advisors' and clients' databases can be used to make on-the-spot decisions on a wide range of issues. Despite the widespread availability of Area Aid land parcel maps a limitation to this application in Ireland is the general absence of soil and other land resource information at farm level.
(b)   Compared with earlier attempts, modern land drainage design is based on rigorous mathematical and physical principles. Drain size, depth and spacing are derived from a combination of site data (geology, topography, rainfall, soil), D'Arcy's Law and nomograms which obviate the calculation of cumbersome mathematical

formulae. Summary details and some examples are given by Mulqueen *et al.* (1999).

(c) An almost instantaneous transfer of information with modern technology is exemplified in such new developments as: (1) harvesting of milled peat by bulk density/moisture content information rather than a pre-set depth method (Ward & Holden, 1998) and (2) in "precision agriculture" where crop (grain, grass) yield variation within a field can be measured swath by swath (by "yieldmeter") and the resulting yield map used to adjust future management (Bailey,1999). The aim of this technology is to reduce tillage, fertiliser and agrochemical costs and to promote the concept of "farming by the soil rather than by the field". In both examples onboard DGPS equipment and adequate computing facilities are essential.

Agricultural Inputs Recommendations

Appraisals under this heading are occasionally referred to as Nutrient Management Plans in which the land unit is generally a catchment or sub-catchment of a river. Recent examples in this country range from tributary catchments of the River Erne, to the Bellsgrove mini-catchment, Co. Cavan (Kirk McClure Morton, 1998). The plans involve an assessment of the soil, hydrological, climatological and topographical conditions of the area with a view to devising a sensible strategy for the timing and rates of application of nutrients to land The main focus is on land-spreading of farm wastes from pig and poultry units, with the desired intention of import substitution and prevention of water pollution. Such plans are now integral parts of Environmental Impact Statements and Integrated Pollution Control licences. At farm level the establishment of a *cordon sanitaire* (buffer zone), the choice of machinery and provision of winter storage facilities may be part of the plans. It is increasingly recognised that substantial variability, both vertically and horizontally, is an inherent feature of all resources, even at field level, and that a line on a map does not indicate an abrupt boundary. A recent seminar drew attention to the need to regard groundwater and surface water as a combined resource (Anon. 1999).

Farming Systems Analysis

Systems Analysis is a blurred version of land evaluation at farm level, the main aim of which is to match soils, crops and livestock. Hence it involves a detailed knowledge of soils and landforms (or Land Utilisation Types) within the farm. It is most useful in mixed farming systems where alternative crops, stocking rates, machinery and other factors can be interchanged. The concept of stocking rate/farm management is currently being applied on a broad scale, based on the perceived state of habitats, in the hill and mountain landscape. Another item of current interest is the assessment of soil-climate-animal requirements which would permit a longer grazing season and associated reduced fodder conservation and housing needs. Care must be taken lest the solution to one problem causes another; the "umbilical" system of slurry spreading mitigates soil compaction but increases the risk of runoff if used indiscriminately in wet weather.

Environmental Impact Analysis

Environmental impact studies and assessments are now commonplace requirements for licensing of large pig, poultry and other farming activities as well as for other land uses, such as motorways, landfill sites, mines, wind farms and fish farms. They are aimed at controlling possible occurrences of air, soil and water pollution. From an agricultural view point, a major focus is on BOD, N and P, and the likely consequences for ground and surface waters (Sherwood,1992). Information from a very wide range of sources must be collated and predictions made on various scenarios (Bradley *et al.*, 1991). The "Advice Notes on Current Practice" published by the Environmental Protection Agency (EPA, 1995b) include the following: (1) *on air*: "The general climate of the site is described as accurately as possible using existing climatic data..." there is an interpretation of the implications of the general climatic conditions for the behaviour of air on and over the development site; (2) *on soils* (and geology): "Suitability/capability classifications or ratings are drawn up for the appropriate land use, and the criteria used in compiling such ratings are specified"; (3) *on water*: "Maps, diagrams and sections illustrating the location and extent of all surface water bodies and aquifers on, or adjacent to, the site are included"; and (4) *on flora*: "It is essential to outline the plants that create or define the habitat...Adequate description may involve the measurement of abundance as well as presence. Such data are collected by standardised and reproducible methods, referenced if possible...".

Monitoring Land Resources Development
Reasons for monitoring land resources development may vary from academic enquiry into the implications of changing land use, to establishing compliance with rules and regulations. Hence historic/archived data of many sorts may have to be processed and compared with current data. The data can be in the form of satellite imagery (e.g. Stanley *et al.*, 1996), air photos, meteorological/hydrometric charts, and ground-truth data (Guinan *et al.*, 1998). Examples include compliance with set-aside agreements for tilled land, stocking rates of marginal land (Walsh & Collins, 1996, 1998), clear-felling of woodland, developments associated with mines and quarries, road construction and urban expansion.

4.8.3   Concluding comments

The contributing authors have searched the literature in their respective disciplines and have highlighted the more important kinds and sources of information relating to the environment they know best: the Irish environment. In doing so they have emphasised its holocoenotic nature, the interconnectedness between the various spheres and scientific disciplines, and the necessity to continually update the databases. They draw the reader's attention to the sources, scope and reliability of Irish data relating to its geological, topographical, hydrological, pedological and biological resources, and how each of these may have implications for agro-meteorological modelling activities.

Surrogates, proxies and transfer functions
Even though they were referring specifically to soils, the place of surrogates/proxies in landscape studies was described by Hole & Campbell (1985). Pointing out that scientists/surveyors cannot observe all properties at all locations, they state that substitutes can help in defining relationships between one property and another. A common example in soil studies is to use natural vegetation to infer soil drainage characteristics. These authors emphasise that the effectiveness of the concept depends

on the initial accuracy of the definition of the relationship between a property and a surrogate, and thereafter the astute application of the relationship.

Not withstanding the multitude of existing datasets, factual data may not be available on many subjects (usually explained on the basis of cost). Covariance and inter-relationships between properties allow estimates of particular probabilities from one, but more effectively from two, three or more, known variables. These mathematical "translations" have been described, with examples, by Bouma (1989) and Larson & Pierse (1994). In soil science jargon they are known as "pedotransfer functions". Two examples are: estimation of cation exchange capacity (CEC) from clay and organic carbon content, and rooting depth from bulk density, pH and available water capacity. Wagenet *et al.* (1991) have alerted readers to the limitations of static attributes of soil characteristics and properties since they tell little about soil processes that are use-dependent. These authors purport that pedotransfer functions and simulation modelling can bridge the gap between assessments of land characteristics and qualities such as trafficability and water supply capacity.

Map data: its use and abuse
Systematic field surveys of natural resources are mostly conducted for multi-purpose use. The data emanating from such surveys are usually in the form of sets of characteristics for specific locations, the sitings of which, amongst others, may be long-term (synoptic, hydrological), widely scattered (drifting buoys), site specific (bore holes), or taxonomically determined (soil augerings). The diagnostic criteria vary with the subject matter, but are usually chosen to allow subsequent interpretation for various types of application or use. Data processing takes the form of first classifying the recorded sets of characteristics and then constructing maps of appropriate scale and legend. The latter process involves spatial interpolation and/or extrapolation of a form specific to each medium: linear (stream quality, hedgerow composition); land surface-based; (soil type, water table depth); computer-generated (atmospheric pressure, wave characteristics), but greatly influenced by the map scale chosen. While there is always a trade-off between legibility and the amount of detail that can be portrayed on a map, the most useful maps are those that have the smallest internal variability. The greatest loss of information occurs when field data are condensed (generalised) for presentation at a much reduced map scale. However, it often happens that, once generalised, the original data source is ignored in any further processing or interpretation studies. Another common misuse of map data is to photo-reduce the original to such an extent that it becomes partially illegible and is accompanied by an absurdly long and inappropriate legend.

Excessive photo-enlargement of maps, especially reconnaissance-type ones, is also an abuse of data. There are numerous examples of attempts at making general data site-specific; even instrumental errors of graphology are often ignored (a line 0.2 mm thick on a 1:50,000 map represents 10 m on the ground; at 1:10,000,000 it represents 2 km). These and similar "mistakes" can be made when layering map data of incompatible scales in a geographic information system.

Over-interpretation and wrongful reading of the map face, legend or caption are commonly observed. Examples noted by the authors include: 1) reproducing the mean air temperature map of Ireland without the caption wording: "reduced to mean sea level"; 2) extrapolating the grazing season map of Ireland (scale c. 1:2,000,000), compiled from mean rainfall and temperature data only, to sheep grazing on specific mountain slopes; 3) basing individual farm management plans on small scale soil maps (at of scale 1:1,000, 000, the local road "occupies" a zone about 0.5 km wide!).

Spatial Variability
In concluding this chapter the authors wish to draw the attention of the reader to a seminal treatise on spatial variability by Mausbach & Wilding (1991). The work includes discussions on the need to quantify spatial variability, predicting soil variability from landscape models, spatial variability in geologic mapping, statistical methods and procedures for evaluating map data, sampling procedures, and quality control. In their abstract of Chapter 1, Arnold & Wilding state that "one of the continuing challenges for pedologists and allied earth scientists is to develop integrated system models to scale spatial knowledge of soils from micro-samples to pedons, landforms and the pedosphere. Quantification of the magnitude, location and causes of spatial variability is an essential but insufficient ingredient of soil surveys. The final payoff is to communicate this knowledge to user clientele in flexible formats that provide for probability risk assessments and alternative land-use decisions." Such sentiments are equally applicable to all the subjects in this chapter as is their axiom:- "Documenting reality to 'be sure' will always be good business and sound science".

Down the centuries man has acknowledged his dependence on the biological and physical resources of the planet. However, their resilience and sustainability has on occasion been stretched almost to the breaking point, neatly summed up by Bradley in 1935 (quoted by Miller *et al.*, 1975) thus:" The parade of civilisation has marched to the cadence of the resources trinity:- soil, water and climate; history is replete with attempts to march to different drummers ...." It is hoped that the data forms and sources assembled in this chapter will set the score for those who aspire to, and work towards, a holistic and non-reductionist view of the world in which we live. As we enter the third millennium, AD, we could profitably recall the elegant definition of the world, by Pliny, at the beginning of the first, and ask how well do we know our little bit of the planet? *Sacer est, aeternus, immensus, totus in toto, immo vero totum, finitus et infinito similis, omnium rerum certus et similus in certo, extra intra cuncta complexus in se, idemque rerum naturae opus et rerum ipsa natura.* Plinius (AD.23-79) Naturalis Historia (II:I).

References

Aalen, F.H.A., K. Whelan & M. Stout. (1997): *Atlas of the Irish Rural Landscape.* Cork University Press, Cork.

Aldwell, C.R., D.J. Burdon & E.P. Daly. (1975): Ireland's Hidden Resource. *Mining Ireland* **3**: 93-98.

An Foras Forbartha. (1977): *Surface Water Year Book 1975* (2 volumes). An Foras Forbartha, Dublin. [Similar volumes area available for 1976 and 1977.]

An Foras Forbartha. (1977): *The South Eastern Water Resource Region. Summary of Hydrometric Records.* An Foras Forbartha, Dublin. [Similar reports are available for each region.].

An Foras Forbartha. (1984a): *A Statistical Analysis of River Flows, South-Eastern Water Resource Region.* An Foras Forbartha, Dublin.

An Foras Forbartha. (1984b): *1984 Drought River Flows. A Comparison with Other Years.* An Foras Forbartha., Dublin.

An Foras Forbartha. (1985): *Water Quality Plan for the River Nore Catchment. Southeast Regional Development Organisation.* An Foras Forbartha, Dublin.

Anon. (1996): *Rural Environment Protection Scheme-Agri-Environmental Specifications.* Department of Agriculture, Food and Forestry, Dublin 2.

Anon. (1999): *Surface water and groundwater: a combined resource.* Proceedings of the 19th Annual Groundwater seminar. International Assocation of Hydrogeologists (Irish Group) Ballymore Eustace, Co. Kildare.

Arnold, R.W. & L.P. Wilding. (1991): The need to quantify spatial variability Chapter 1 in: Mausbach, M.J. & L.P. Wilding, L.P., editors, *Spatial Variabilities of Soils and Landforms.* SSSA Special Publication Number 28, Madison Wisconsin, USA.

Bailey, J.S. (1999): Modelling sward productivity using plant tissue analyses. pp. 21-22 In: P. O'Kiely, T. Storey & J.F. Collins. (Editors): *Agricultural Research Forum.* UCD, Dublin.

Beek, K.J. (1978) *Land Evaluation for Agricultural Development.* INLI Publication 23, International Institute of Land Reclamation and Improvement, Wageningen.

Betts, N.L. (1982) Climate. Chapter 1 In: Cruickshank, J.G. & D.N. Wilcock, (Editors): *Northern Ireland Environment and Natural Resources.* QUB and NUU.

Bouma, J. (1989): Using soil survey data for quantitative land evaluation. *Advances in Soil Science* **9**: 177-213.

Boyle, G.M., E.P. Farrell & T. Cummins. (1997): *Intensive Monitoring Network:Ireland.* Forem 2 Final Report:Project No. 9360 IR 0030, UCD. Dublin.

Bradley, K., C. Skehan & Walsh. (Editors). (1991): *Environmental Impact Assessment: A technical approach.* DTPS Ltd., Dublin.

Brady Shipman Martin.(1997): *Coastal Zone Management, A Draft Policy for Ireland.* Government of Ireland, Dublin.

Braun-Blanquet, J. & R. Tüxen. (1952): Irisches Pflanzengesellschaften. *Veröffentlichungen des Geobotanischen Institutes Rübel in Zürich* 25, 224-415.

Brereton, A.J. & T. Keane. (1982): The effect of water on grassland productivity in Ireland. *Irish Journal of Agricultural Research* **21**: 227-248.

Brereton, A.J. (1972): The influence of soil type on crop production. *Biatas* **25**: 444-448.

CEC (1985) *Soil Map of the European Communities* 1:1,000,000 DGVI, Office for the Official Publications of the European Communities, Luxemburg.

Chidley, T.R.E., J. Elgy & J. Antoine/ (1993): *Computerized Systems of Land Resources Appraisal for Agricultural Development.* Land and Water Development Division, F.A.O. Rome.

Clabby, K.J., J. Lucey & M.L. McGarrigle. (1982): *The National Survey of Irish Rivers. River Quality Investigations: Biological. Results of the 1980 and 1981 Investigations.* An Foras Forbartha, Dublin.

Clabby, K.J., J. Lucey & M.L. McGarrigle. (1995): *Interim Report on The Biological Survey of River Quality, Results of the 1994 Investigations.* Environmental Protection Agency, Dublin.

Clabby, K.J., J. Lucey, M.L. McGarrigle, J.J. Bowman, P.J. Flanagan & P.F. Toner. (1992): *Water Quality in Ireland 1987-1990. Part One: General Assessment.* Environmental Research Unit, Dublin.

Collins, J. F. & T. Cummins. (1996): *Agroclimatic Atlas of Ireland.* AGMET Group Dublin.

Collins, J.F. (1981): Relationships between Drift Geology, Soils and Agriculture in Ireland-Current View, 1980. *Agricultural Record (250 Years of the R.D.S Special Commemorative Issue)* 1-8

Connaughton, M. J. (1973): *The Grass Growing Season in Ireland.* Agrometeorological Memorandum No. 5. Irish Meteorological Service, Dublin.

Conry, M.J. & A. Hegarty. (1997): Effects of depth phases of the Clonroche soil series and farming skill on spring barley yields. *Irish Geography* **30**: 1-8.

Conry, M.J. (1985): Comparison of grain yield and quality of winter and spring barleys grown on four contrasting soils in Ireland. *Irish Journal of Agricultural Research* **24**: 201-212

Conry, M.J. (1996): Effect of fertilizer-N on grain yield and quality of spring barley grown on five contrasting soils. *Biology and Environment* **97**: 185-196.

Convery, F. & J. Feehan. (Editors). (1995): *Assessing Sustainability in Ireland.* Conference Proceedings, The Environmental Institute UCD.

Corish, P.N. (1975): *Methods of River Flow Gauging.* Hydrometric Section, Office of Public Works.

Crawford, I., A. Bleasdale & J. Conaghan. (1996): *Biomar Survey of Irish Machair Sites: 1. Site Information; 2. Plant Communities.* Irish Wildlife Manuals, 3 & 4. Dúchas, The Heritage Service, Dublin.

Cross, J. (1992): The distribution, character and conservation of woodlands on esker ridges in Ireland. *Proceedings of the Royal Irish Academy* **92B**: 1-19.

Cross, J. (1998): An outline and map of the potential natural vegetation of Ireland. *Applied Vegetation Science* **1**: 241-252.

Cross, J.R. (1990): *The Raised Bogs of Ireland: their Ecology, Status and Conservation.* The Stationery Office, Dublin.

Cruickshank, J. G. (1997): *Soil and Environment: Northern Ireland.* Department of Agriculture Northern Ireland /The Queen's University, Belfast.

Cruickshank, J.G. (1984): Soils and Bio-geography in Ireland, 1934-1984. Chap VI In G.L. Herries Davies (Editor): *Irish Geography.* The Geographical Society of Ireland, Dublin.

Cruickshank, M.M. & R.W. Tomlinson. (1990): Peatlands in Northern Ireland: inventory and prospect. *Irish Geography* **23**: 17-30.

Cruickshank, M.M., R.W. Tomlinson, P.M. Devine & R. Milne, (1998): Carbon in the vegetation and soils of Northern Ireland. *Biology and Environment*, **98**: 9-21.

Cruikshank, M.M. & R.W. Tomlinson. (1996): Application of CORINE land cover methodology to the U.K.-some issues raised from Northern Ireland. *Global Ecology and Biogeography Letters* **5**: 235-248.

Cullinan, E F. & M. Bulfin (1996): Development of an Indicative Forest Strategy with special reference to Co. Clare. *Irish Forestry* **53**: 69-77.

Cummins, T. & D.P. Whelan. (1996): *Woodland Establishment and Management for the Timber Grower*. The Irish Timber Growers' Association, Dublin.

Curtis, T.G.F. & M.J. Sheehy Skeffington. (1998): The salt marshes of Ireland: an inventory and account of their geographical variation. *Biology and Environment* **98B**: 87-104.

Curtis, T.G.F. (1991): A site inventory of the sandy coasts of Ireland-their types and distribution. pp. 6-17. In M.B. Quigley. (Editor): *A Guide to the Sand Dunes of Ireland*. European Union for Dune Conservation and Coastal Management, Dublin.

Daly, D. & W.P. Warren. (1998): Mapping groundwater vulnerability: the Irish perspective. In: Robins, N. S. (Editor): *Groundwater Pollution, Aquifer Recharge and Vulnerability*. Geological Society London, Special Publication 130.

Daly, D. (1998): *Landspreading of organic wastes and groundwater protection*. Seminar International Inititude of Hydrogeologists (Irish Group), Portlaoise.

Daly, D., R. Thorn & H. Henry. (1993): *Septic Tank Systems and Groundwater in Ireland*. RS 93/1, Geological Survey of Ireland, Dublin.

Department of Industry and Energy. (1986): *The Development of Small Scale Hydro-schemes. Part 1 Rapid Assessment*. Department of Industry and Energy, Dublin.

Department of Local Government/An Foras Forbartha. (1974): *Water Resources in Ireland, A Preliminary Statement of the Present Position*. An Foras Forbartha., Dublin.

Devillers, P., J. Devillers-Terschuren & J.P. Ledant. (1991): *CORINE Biotopes Manual: Habitats of the European Community. Data specifications, part 2*. Commission of the European Communities, Luxembourg.

Egan, N., J.F. Collins & M. Walsh. (1996): Teagasc Hill Sheep Farm 1, Physical and Soil Resources. *Farm and Food Research* **6**: 12-15.

EHS. (1996): *River Water Quality in Northern Ireland 1995*. Environment and Heritage Service, Belfast.

ENFO. (1990): *Measurement of River Water Quality*. Briefing Sheet 12. Environmental Information Service, Dublin.

Environment Service. (1994): *River and Estuary Quality*. Report of the 1991 Survey, Environment Service, Belfast.

Environment Service. (1996): *Report for 1993-1995 Period*. Environment Service, Belfast.

EPA. (1995): *Hydrological Data. Map 1, Hydrometric Network, Water Level Recorders*. Environmental Protection Agency, Wexford.

EPA. (1995a): *Draft guidelines on the information to be contained in environmental impact statements*. Environmental Protection Agency, Dublin.

EPA. (1995b): *Advice Notes on Current Practice*. Environmental Protection Agency, Dublin.

EPA. (1999): *Water Quality in Ireland 1995-1997*. Environmental Protection Agency, Wexford.

ERU. (1989): *A Statistical Analysis of River Flows*. The Eastern Water Resource Region. Environmental Research Unit, An Foras Forbartha, Dublin.

FAO. (1975): FAO-UNESCO 1974 *Soil Map of the World at 1:5,000,000 scale*. UNESCO, Paris.

Farrell, E. P. (1995): Atmospheric deposition in marine environments and its impact on terrestrial ecosystems. *Water, Air and Soil Pollution* **85**: 1653-1658

Feehan, J. & G. O'Donovan. (1996): *The Bogs of Ireland*. The Environmental Institute, University College, Dublin.

Ferguson, D.K. & V. Westhoff. (1987): An account of the flora and vegetation of Derryclare Wood, Connemara (Co. Galway), Western Ireland. *Proceedings of the Koninklijke Nederlandse Akademie van Wetenscha ppen* **90C**: 139-172.

Flanagan, P.J & P.M. Larkin. (1992): *Water Quality in Ireland (1987-1990). Part Two; River Quality Data*. ERU, Dublin.

Flanagan, P.J. & P.F Toner. (1975): *A Preliminary Survey of Irish Lakes*. An Foras Forbartha, Dublin.

Flanagan, P.J. & P.F. Toner. (1972): *The National Survey of Irish Rivers, A Report on Water Quality*. An Foras Forbartha, Dublin.

Flanagan, P.J. (1972): *The National Survey of Irish Rivers: A Second Report on Water Quality*. An Foras Forbartha, Dublin.

Galvin, L.F. (1981): Land Drainage Survey III : Maps of Ireland showing the general distribution of drainage problems and of drainage schemes. *Irish Journal of Agricultural Research* **10**: 213-221.

Gardiner, M. J. & T. Radford. (1980): *Soil Associations of Ireland and their Land-Use Potential*. Soil Survey Bulletin No.36. An Foras Talúntais, Dublin, pp.143.

Gardiner, M. J. (1981) The value of the soil survey in regional development. *Agricultural Record* **41**: 25-34.

Gardiner, M. J. (1982): Use of regional and global soils data for climate modelling. pp 361-393 In: P. S. Eagleson. (Editor): *Land Surface Processes in Atmospheric General Circulation Models*. Cambridge University Press.

Guinan, L., M. Walsh, J.F. Collins & D. Nixon. (1998): Evaluation of the impact of livestock on the hill environment using a digital photogrammetric system. pp 163-164. In P. O'Kiely, T. Storey & J.F. Collins. (Editors): *Proceedings of Agricultural Research Forum*. Dublin.

Hammond, R .F. (1981): *The Peatlands of Ireland*. Soil Survey Bulletin No.35. An Foras Talúntais, Dublin.

Hardisty, J. (1990): *The British Seas*. Routledge, London.

Hayes, R.J. (1965): *Manuscript Sources for the History of Irish Civilisation*. (Unpublished, photocopies in all main libraries).

Heery, S. (1991): The plant communities of the grazed and mown grasslands of the River Shannon callows. *Proceedings of the Royal Irish Academy* **91B**: 1-19.

Heineke, H.J., W. Eckelmann, A.J. Thomasson, R.J.A. Jones, L. Montanarella & B. Buckley. (Editors). (1998): *Land Information Systems: Developments for Planning the Sustainable Use of Land Resources*. European Soil Bureau, Joint Research Centre, I-21020, Ispra, Italy.

Herries Davies, G.L. (1995): *North from the Hook-150 years of the Geological Survey of Ireland*. Geological Survey of Ireland, Dublin.

Hole, F.D. & J.B. Campbell. (1985): *Soil Landscape Analysis*. Routledge & Kegan Paul, London.

Horner, A.A., J.A. Walsh & J.A. Williams. (1984): *Agriculture in Ireland, a Census Atlas*. Department of Geography, University College, Dublin.

Hussain, S. M., G.W. Smillie & J.F. Collins. (1985): Laboratory studies of crust development in Irish and Iraqi soils Soil & Tillage Research **5**: 33-53.

Keane, T. (Editor). (1986): *Climate, Weather and Irish Agriculture*, AGMET Group, Dublin.

IHP. (1982): *Hydrology in Ireland*. A Contribution to the UNESCO International Hydrological Programme. Irish National Committee for the International Hydrological Programme, United Nations (UNESCO) Dublin.

Jaritz, G. & J. Lee. (1968): *The Soils and Vegetation of the Economic Test Farm, Drumboylan, Co. Roscommon*. Soil Survey Bulletin 19. An Foras Taluntais, Dublin.

Keane, T. (1988): Features of the Irish climate of importance to agriculture: comparison with neighbouring Europe. In : Keane, T. (Editor): *Proceedings of Conference on Weather and Climate*. Agmet Group, Dublin.

Kelly, D. & S.F. Iremonger. (1997): Irish wetland woods: the plant communities and their ecology. Biology and Environment **97B**: 1-32.

Kilroe, J. R. (1907): *A Description of the Soil-Geology of Ireland*. Alex Thom, Dublin.

Kirk McClure Morton. (1997): *Proposals for a Water Quality Management Strategy for the Foyle Catchment and Lough Foyle (Executive Summary)*. Environment and Heritage Service and Department of the Environment and Local Government.

Kirk McClure Morton. (1998): *Lough Derg & Lough Ree Catchment Monitoring and Management System, Interim Report*. Monksland Civic Offices, Monksland, Athlone, Co. Roscommon.

Klingebiel, A.A & P.H. Montgomery. (1961): *Land Capability Classification*. USDA Agricultural Handbook 210, Soil Conservation Service, USDA, Washington, DC.

Lafferty, S., P. Commins & J.A. Walsh. (1999): *Irish Agriculture in Transition; a Census Atlas of Agriculture in the Republic of Ireland*. Department of Geography, NUI, Maynooth.

Larney, F. J. (1985): *Effects of soil physical conditions, deep loosening and seedbed cultivations on growth and yield of sugar beet*. Ph. D. Thesis. The National University of Ireland, Dublin.

Larson, W.E. & F.J. Pierce, F.J. (1994): The dynamics of soil quality as a measure of sustainable management. In: Doran, J. W., J.A.E. Molina and R. F. Harris. (Editors): *Defining Soil Quality For a Sustainable Environment*. SSSA Special Publication Number 35. SSSA Inc., Madison , Wisconsin, USA.

Lee, J. & Ryan, P. (1966): Soil survey interpretation for crop productivity determinations - 1:Relation between soil series and sugar-beet yields. *Irish Journal of Agricultural Research* **5**: 237-248.

Lee, J. & S. Diamond. (1972): *The Potential of Irish Land for Livestock Production*. Soil Survey Bulletin No. 26. An Foras Talúntais, Dublin.

Lee, J. (1986): European land 1: How good is Ireland's share? *Farm and Food Research* **17**: 4-6.

Lennox, L.J. & P.F. Toner. (1980): *The National Survey of Irish Rivers. A Third Report on Water Quality*. An Foras Forbartha, Dublin.

MacCarthaigh, M. (1992): *Assessment and Forecasting of Drought Flow Conditions in Irish Rivers*. Paper presented to Water & Environmental Section, Institution of Engineers of Ireland, Dublin.

MacCarthaigh, M. (1995): *Hydrological Data. A listing of water level recorders and summary statistics at selected gauging stations*. Environmental Protection Agency, Wexford.

MacCarthaigh, M. (1996): *An Assessment of the 1995 Drought. Including a Comparison with other known Drought Years.* Environmental Protection Agency, Wexford.

MacCarthaigh, M. (1999): *Surface Water Resources in Ireland.* A paper presented to the Institute of Engineers of Ireland as part of Module 1 of the IEI Water Course, IEI, Dublin.

Maloney, M. (1994): *Agriculture and the Environment.* Royal Dublin Society.

Marine Institute (1996): *Towards a Marine Policy for Ireland.* Proceedings of the Consultative Process. Marine Institute, Dublin.

Martin, J.V. & C. Cunnane. (1977): Analysis and prediction of low flows and drought volumes for selected Irish rivers. *Transactions of the Institute of Engineers of Ireland* **101**: 21-28.

Martin, J.V. & C. Cunnane. (1994): River flow in Ireland-characteristics and measurements. In: T. Keane & E. Daly. (Editors): *Proceedings of AGMET and Royal Meteorological Society Conference, The Balance of Water-Present and Future..* AGMET, Dublin.

Martin, J.V. (1992): *Floods and Droughts 1989-1991 1. Analysis of Floods and Drought Volumes.* Paper presented to Water & Environmental Section, Institution of Engineers of Ireland.

Mausbach, M. J. & L.P. Wilding. (1991): *Spatial Variabilities of Soils and Landforms.* SSSA Special Publication Number 28, SSSA Inc., Madison, Wisconsin, USA.

McCumiskey, L.M. (1982): A Review of Ireland's Water Resources-Rivers and Lakes. *Irish Journal of Environmental Science* 1: 1-13.

McCumiskey, L.M. (1991): *Water in Ireland.* Environmental Research Unit. Dublin.

McEntee, M.A. (1979): *Climate as a factor in grass growth.* Grass Production Seminar pp. 1-13. An Foras Talúntais, Johnstown Castle, Wexford.

McIsaac, G. F. & M. Brun. (1999): Natural environment and human culture: defining terms and understanding world views. *Journal of Environmental Quality* **28**: 1-10.

Meyer-Roux, J. & J. Montanarella. (1998): The European Soil Bureau. pp.3-10 In: Heineke H.J., W. Eckelmann, A.J. Thomasson, R.J.A. Jones, L. Montanarella & B. Buckley. (Editors): *Land Information Systems Developments for Planning the Sustainable Use of Land Resources.* European Soil Bureau Research report No. 4, EUR 17729. EN. Luxembourg.

Miller, F.P., W.D. Rasmussen & L.D. Myer. (1975): Historical perspectives of soil erosion in the United States. Chapter 3 In: Follet R.F. & B.A. Stewart. (Editors): *Soil Erosion and Crop Productivity.* A.S.A. Madison, Wisconsin, U.S.A.

Moles, R. & I. Travers. (1983): *Mullach Mór, Co. Clare, Ireland: Habitat Map.* Privately published.

Mooney, E.P. & M. O'Connel. (1990): The phytosociology and ecology of the aquatic and wetland plant communities of the Lower Corrib Basin, Co. Galway. *Proceedings of the Royal Irish Academy* **90B**: 57-97.

Moore, J.J. & H. O'Reilly. (1977): Saltmarsh: vegetation pattern and trends. pp. 83-87. In: D.W. Jeffrey. (Editor): *North Bull Island Dublin Bay, a Modern Coastal Natural History.* Royal Dublin Society, Dublin.

Moore, J.J. (1960): A re-survey of the vegetation of the district lying south of Dublin (1905-1956). *Proceedings of the Royal Irish Academy* **61B**: 1-36.

Moore, J.J. (1968): A classification of the bogs and wet heaths of northern Europe. pp. 306-320. In R. Tüxen. (Editor): *Pflanzensoziologische Systematik*. Dr W. Junk, The Hague.

Mueller-Dombois, D. & H. Ellenberg. (1974): *Aims and Methods of Vegetation Ecology*. John Wiley, New York.

Mulqueen, J., M. Rodgers, E. Hendrick, M. Keane & R. McCarthy. (1999): *Forest Drainage Engineering, A Design Manual*. COFORD, Dublin.

NERC. (1975): *Flood Studies*. Natural Environment Research Council, London.

Ní Lamhna, E. (1982): The vegetation of saltmarshes and sand-dunes at Malahide Island, County Dublin. pp. 111-129. In: J. White. (Editor): *Studies on Irish Vegetation*. Royal Dublin Society, Dublin.

O'Connell, M., J.B. Ryan & B.A. MacGowan. (1984): Wetland communities in Ireland: a phytosociological review. pp. 303-364. In: P.D. Moore. (Editor): *European Mires*. Academic Press, London.

Ó Críodáin, C. & G.J. Doyle. (1994): An overview of Irish small-sedge vegetation: syntaxonomy and a key to communities belonging to the Scheuchzerio-Caricetea nigrae (Nordh. 1936) Tx. 1937. *Biology and Environment* **94B**: 127-144.

OECD. (1982): *Eutrophication of Waters*. Monitoring, Assessment and Control. Organisation for Economic Cooperation and Development, Paris.

Open University. (1989): *Waves, Tides and Shallow-Water Processes*. Pergamon Press, Oxford.

OPW. (1987): *Records of Flows, Brownsbarn Gauging Station* (1506, R. Nore). Office of Public Works, Dublin (unpublished).

OSI. (1958): *Ireland: Rivers and their Catchment Basins*. Map. Ordnance Survey, Dublin.

OSI. (1996): *The Irish Grid; A description of the Co-ordinate Reference System used in Ireland*. Ordnance Surveyof Ireland, Dublin.

OSI. (1998): *Catalogue - Products and Services*. Ordnance Survey of Ireland, Dublin.

OSNI. (1994): *Maps and Services Catalogue*. Ordnance Survey of Northern Ireland.

O'Sullivan, A. M. (1986): *Botanical Survey-Field Station, Kilmaley, Co Clare*. Irish Vegetation Studies 4, An Foras Talúntais, Dublin

O'Sullivan, A.M. (1965): *A Phytosociological Survey of Irish Lowland Meadows and Pastures*. Ph.D. Thesis, National University of Ireland, Dublin.

O'Sullivan, A.M. (1982): The lowland grasslands of Ireland. pp. 131-142 in J. White, editor, *Studies on Irish Vegetation*. Royal Dublin Society, Dublin.

O'Sullivan,G. (Editor). (1994): *CORINE Land cover project (Ireland) Project Report*. Ordnance Survey of Ireland and Ordnance Survey Northern Ireland for CEC, DGXI and DGIV.

Ozenda, P. (1979): *Vegetation Map of the Council of Europe Member States*. Council of Europe, Strasbourg.

Perring, F.H. & S.M. Walters. (1962): *Atlas of the British Flora*. Thomas Nelson, London.

Pethybridge, G.H.& R.L. Praeger. (1905): The vegetation of the district lying south of Dublin. *Proceedings of the Royal Irish Academy* **25B**: 124-180.

Praeger, R.L. (1907): Contributions to the natural history of Lambay, Co. Dublin. Phanerogams and vascular cryptogams. *Irish Naturalist* **16**: 90-99.

Praeger, R.L. (1911): Clare Island Survey. Part 10. Phanerogamia and Pteridophyta. *Proceedings of the Royal Irish Academy* **31**: 1-112.

RIA. (1979): *Atlas of Ireland*. Royal Irish Academy, Dublin.

Rodwell, J. (Editor). (1991-'99): *British Plant Communities, Vols 1-V*. Cambridge University Press, Cambridge.

Schaminée, J.H.J., A.H.F. Stortelder & E.J. Weede. (1995-'99): *Die Vegetatie van Nederland*, Vols 1-5. Opulus Press, Uppsala.

Sherwood, M. (1992): *Weather, Soils and Pollution from Agriculture*. AGMET Group Dublin.

Shine, H. (1987): *Flood warning system for Kilkenny City*. Proceedings of a Seminar on Bridge Collapse Causes, Consequences & Remedial Measures. An Foras Forbartha, Dublin, pp. 215-224.

Soil Survey Staff. (1993): *Soil Survey Manual*. Handbook No. 18 USDA Washington DC.

Stanley, B., Dunne, S. & Keane, M. (1996): Forest condition assessments and other applications of colour infrared (CIR) aerial photography in Ireland. *Irish Forestry* **53**:19-27.

Storie, R.E. (1933): *An Index for Rating the Agricultural Value of Soils* (revised 1937). University of California Agricultural Experiment Station Bulletin 556, Berkeley, California.

Sweeney, J. (Editor). (1997): *Global Change and the Irish Environment*. Proceedings of the 1st National IGBP Symposium, Royal Irish Academy, Dublin.

Sweeney, J.C. (1989): *The Irish Sea: a Resource at Risk*. Special Publications No. 3, Geographical Society of Ireland, Dublin.

Taylor, A.G., J.E. Gordon & M.B. Usher. (Editors). (1996): *Soils, Sustainability and the Natural Heritage*. HMSO, Edinburgh.

Tomlinson, R.W. (1984): Mapping peatland vegetation from readily available air-photographs: a field test from Northern Ireland. *Irish Geography* **17**: 65-83.

Toner, P.E & A.J. O'Sullivan. (1977): *Water Pollution in Ireland*. National Science Council.

Toner, P.F., K.J. Clabby, J.J. Bowman & M.L. McGarrigle. (1986): *Water Quality in Ireland. The Current Position. Part One: General Assessment*. An Foras Forbartha, Dublin.

Van Diepen, C.A., H. van Keulen, J. Wolf & J.A. Berkhout. (1991): Land evaluation: from Intuition to Quantification. Stewart, B.A., editor, *Advances in Soil Science*, **15A**. Springer-Verlag, New York.

Wagenet, R.J., J. Bouma & R.B. Grossman. (1991): Minimum data sets for use of soil survey information in soil interpretive models. In: Mausbach, M. J. & L.J. Wilding. (Editors): *Spatial Variabilities of Soils and Landforms*. SSSA Special Publication Number 28. Madison, Wisconsin,U.S.A.

Walsh, M. (1996): Is sheep overgrazing just a climatic effect? In: Coulter, B., N. Culleton & W. Murphy. (Editors): *Proceedings of Agmet Conference–Climate, Weather and the Environment*, Teagasc, Wexford.

Walsh, M., & J.F. Collins. (1998): Teagasc Hill Sheep Farm, Leenaun: - Changes in vegetation 1995-1997 by physiographic unit. pp. 161-162. In: O'Kiely, P., T. Storey & J.F. Collins. (Editors): *Proceedings of Agricultural Research Forum*. Dublin

Ward , S. M. & Holden, N. M. (1998): Precision peat production. *Precision Agriculture, Proceeding of the 4th International Conference, 19-22 July 1998, Saint Paul, Minnesota, USA.* Edited by P C Robert, R. H. Rust and W E Larson. ASA/CSSA/SSSA, Madison WI. p 937-942.

Webb, D.A. (1947): The vegetation of Carrowkeel, a limestone hill in north-west Ireland. *Journal of Ecology* 35: 105-129.

Webb, D.A. (1978): Flora Europaea: a retrospect. *Taxon* 27: 3-14.

Whitby, M.C. (Editor). (1992): *Land use change: the causes and consequences.* ITE Symposium no. 27 HMSO, London.

White, J. & G. Doyle. (1982): The vegetation of Ireland-a catalogue raisonné. pp. 289-368. In: J. White (Editor): *Studies on Irish Vegetation.* Royal Dublin Society, Dublin.

White, J. (1982a): A history of Irish vegetation studies. pp. 15-42. In: J. White, (Editor): *Studies on Irish Vegetation.* Royal Dublin Society, Dublin.

White, J. (1982b): *Studies on Irish Vegetation.* Royal Dublin Society, Dublin.

White, J. (1982c): The Plantago sward, Plantaginetum coronopodo-maritimi. pp. 105-110 In: J. White. (Editor): *Studies on Irish Vegetation.* Royal Dublin Society, Dublin.

White, J. (1985): The census of plants in vegetation. pp. 33-88. In J. White. (Editor): *The Population Structure of Vegetation.* Dr W. Junk, Dordrecht.

Wilcock, D.N. (1997): Rivers, drainage basins and soils. In: J. G. Cruickshank, (Editor): *Soil and Environment: Northern Ireland.* Department of Agriculture Northern Ireland, The Queen's University, Belfast.

Wright, G.R. (1976): Groundwater for Industry. *Technology Ireland* 8: 12-16

Wymer, E.D. (1984): *The Phytosociology of Irish Saltmarsh Vegetation.* M.Sc. Thesis, National University of Ireland, Dublin.

Addresses/contacts

Mapping services
European Environment Agency, Kogens Nytorv 6, DK-1050 Copenhagen.
European Air Services, The Stables, Portmarnock, Co. Dublin.
ERA-MAPTEC, Ltd., Satellite Data, 36 Dame St., Dublin 2
Geological Survey of Ireland , Beggars' Bush, Dublin 4.
Ordnance Survey Office, Phoenix Park , Dublin 8
Ordnance Survey of Northern Ireland, Colby House Stranmillis Court, Belfast BT9 5BJ
Spectral Signatures Ltd., Roebuck Castle UCD, Dublin 4

Groundwater services
Groundwater Section, Geological Survey of Ireland, Beggars Bush, Haddington Road, Dublin 4. Tel: +353 1 6707444. Fax: +353 1 6681782. www.gsi.ie
Geological Survey of Northern Ireland, 20 College Gardens, Belfast BT9 6BS Tel: +00 44 02890 666595   Fax: +00 44 02890 662835
Environmental Protection Agency, Dublin Regional Inspectorate, St. Martins House, Waterloo Road, Dublin 4. Tel: +353 1 6602511. Fax: +353 1 6680009. www.epa.ie

Surface water information sources
Anon. (1977) Water Wells. Information Circular, Geological Survey of Ireland, Dublin.

County Tipperary (South Riding) Groundwater Protection Scheme 1995, Revised 1998.
County Meath Groundwater Protection Scheme 1996.
County Offaly Groundwater Protection Scheme 1998.
County Waterford Groundwater Protection Scheme 1998.
County Limerick Groundwater Protection Scheme 1996.
County Wicklow Groundwater Protection Scheme (Draft) 1998.
County Claire Groundwater Protection Scheme 2000.
County Laois Groundwater Protection Scheme 2000.
South County Cork Groundwater Protection Scheme 2000.
Hydrogeological Map of Northern Ireland (1:250,000) (1994) British Geological Survey & Department of the Environment for Northern Ireland, Environment Service.
Groundwater Vulnerability Map of Northern Ireland (1:250,000) (1994) British Geological Survey & Department of the Environment for Northern Ireland, Environment Service
Soil Survey and Land Research Centre & Department of Agriculture for Northern Ireland
IHME (1980) International Hydrogeological Map of Europe, 1976.
NERDO (1981) North East Regional Development Organisation. An Foras Forbartha & Geological Survey of Ireland.

Marine services
An Bord Iascaigh Mhara, Crofton Road, Dun Laoghaire, Co. Dublin, Ireland.
British Oceanographic Data Centre, Proudman Oceanographic Laboratory,
Bidston Observatory, Birkenhead, Merseyside L43 7RA, United Kingdom.
Department of the Marine, Leeson Lane, Dublin 2, Ireland
Imray, Laurie, Norie & Wilson Ltd., Wych House, The Broadway, St. Ives, Huntingdon PE17 4BT, United Kingdom.
Irish Marine Institute, 80 Harcourt Street, Dublin 2, Ireland
Martin Ryan Marine Institute, National University of Ireland, Galway, Ireland
Met Éireann, Glasnevin Hill, Dublin 9, Ireland.
UK Hydrographic Office, Admiralty Way, Taunton, Somerset TA1 2DN, United Kingdom (http://www.hydro.gov.uk/home.htm)

Soil services
Agricultural and Environmental Science Division, Department of Agriculture Northern Ireland, Newforge Lane, Belfast BT9 5PX.
Agricultural and Environmental Science Department, The Queen's University, Newforge Lane, Belfast, BT9 5PX.
Teagasc (formerly An Foras Talúntais), H.Q. and library, 19 Sandymount Ave. Dublin 4.
Teagasc (formerly An Foras Taluntais), Soils and Grassland Division, Johnstown Castle, Wexford.
Coillte Teoranta, The Irish Forestry Board, Soils Laboratory, Newtownmountkennedy, Co. Wicklow.
Department of Agriculture and Food, Agriculture House, Kildare St., Dublin 2.
Independent Analytical Services (IAS), Bagenalstown, Co. Carlow.

Soil Science Laboratories, Agriculture Building, University College Dublin, Belfield, Dublin 4.

Outside Ireland

Food and Agriculture Organisation of the United Nations (FAO), Viale della Terme di Caracalla, Rome, Italy.

International Soil Reference and Information Centre (ISRIC), Agricultural University, Wageningen, The Netherlands

Office for Official Publications of the European Communities, Luxemburg.

The European Soil Bureau, Joint Research Centre, I-21020, Ispra, Italy.

United States Department of Agriculture (USDA), Superintendent of Documents, U.S. Government Printing Office, Washington, DC 20402.

Chapter 5   A Comparison of Grass Growth Models

A. J. Brereton[1] and E. O'Riordan[2]
1.   *Ecobiops, Oliver Plunkett Chambers, Cork, Ireland*
2.   *Teagasc, Grange Research Centre, Dunsany, Co. Meath, Ireland*

5.1   *Introduction to Grass Models*

In Chapter 1 the fundamental aspects of modelling were described. In this chapter three different approaches to modelling the same system are analysed. All of the models were developed in the context of grass production in agriculture in temperate north-west Europe. In each case temperature and radiation are the driving variables. Grass dry matter mass is the output that is common to all three models. Although they have that much in common, the models are individually quite different. The differences reflect the different application contexts of each. One of the models was developed at the Grassland Research Institute at Hurley in England (Johnson, Ameziane & Thornley 1983; Johnson & Thornley 1983; Johnson & Parsons 1985). It is a mechanistic, dynamic, deterministic model which provides a means to understand grass growth as an integrated system of interacting component processes. The second model was developed at the DLO Institute for Agrobiology and Soil Fertility at Wageningen in The Netherlands (Bouman, Schapendonk, Stol & van Kraalingen 1996). It is also a mechanistic, dynamic, deterministic model but it was not designed as an explanatory model and the number of component processes is restricted. The model was developed for the purpose of monitoring grass production at 10-day intervals throughout the member states of the EU. The third model, developed at the Teagasc research centre at Johnstown Castle in Wexford, Ireland (Brereton, Danielov and Scott 1996) is an empirical model which was developed for the purpose of understanding the behaviour of a grassland farm system in response to variable herbage supply (related to weather variation). In the description presented here the Irish model has been modified in several respects. For convenience the three models are referred to as the English, the Dutch and the Irish model respectively.

Each model is described in a sequence beginning with leaf photosynthesis and ending with grass dry matter production. Sufficient detail is given to enable the reader to gain a general understanding of the construction and functioning of each model. Weather data for five years were used to compare the estimates of grass production from each of the models, and these are compared with measured grass production. The parameters given in the source publications were used in the case of the English and Dutch models. Following the changes made in the Irish model it was re-calibrated using an independent set of weather data rather than in the form of the original publication. Each model was programmed in Visual Basic to read a weather file containing day, month, year, mean air temperature ($^{\circ}$C), rainfall (mm) and duration of bright sunshine (hours) for each day of a year. The three models were run using identical data.

5.2   An outline of the growth of the grass crop

A crop of grass is composed of individual grass plants. Each plant is composed of individual tillers. The tiller is the unit that is most easily recognised on the ground and it may be regarded as the basic unit of the crop. It consists of an underground root system and an over-ground system of leaves and stem. When the tiller is in the reproductive state the stem also carries an inflorescence. Between mid-summer and mid-winter, when tillers are in a vegetative state, the stem is greatly reduced and the tiller mainly consists of leaves. Leaves are formed at the stem apex and as each new leaf emerges into the canopy an older leaf becomes senescent and dies. As a result, the vegetative grass tiller usually bears a constant number of three green leaves. During January, an inflorescence begins to form at the stem apex and new leaf initiation ceases. Leaves that emerge subsequently are the leaves that were initiated but not developed before the inflorescence began to form. During April, May and June the inflorescence completes its development and the stem enlarges and elevates the inflorescence above the leaf canopy. It also elevates the younger leaves so that light conditions within the sward are altered. When flowering is completed in June, the tillers die if they are not harvested, and are replaced by vegetative tillers.

The seasonal transition between the reproductive and vegetative state is associated with a distinct seasonal pattern of growth rates. Growth rates in the April-June period can exceed 100 kg dry matter (dm) ha^{-1} day^{-1}. Later in the summer growth rates are typically 50 kg dm ha^{-1} day^{-1}. This seasonal pattern is of practical significance as it determines the pattern of grassland management on livestock farms. In Spring and early Summer the greater growth rates make it possible to restrict the grazing herd to approximately half of the farm so that the other half can be used for Winter forage production. The greater growth rates required to support the reproductive state appear to be associated with changes in leaf morphology (Stapleton & Jones, 1989) and with changes in light relations within the canopy as the stem enlarges and leaves are elevated (Parsons & Robson, 1982). The English model to be described was constructed specifically for vegetative growth and the Dutch model makes no specific reference to the different states. This leads to the expectation that these models would tend to underestimate growth during the reproductive stage. The present analysis shows this to be the case in the Dutch model but not in the English model. The Irish model incorporates a specific provision for the seasonal change between reproductive and vegetative growth.

Growth represents the balance between the two fundamental processes of photosynthesis, production of carbohydrate substrates, and respiration consumption of carbohydrates resulting in the formation of new plant matter and maintenance of the existing plant. However, growth is not simply the balance between photosynthesis and respiration. In many situations, growth is led by the ability of the plant to use assimilates for new tissue formation, and in this case it is the morphogenetic processes which are limiting plant growth. Growth of the plant is co-limited by both assimilate supply and assimilate use because the two processes are intimately interrelated by several feed-back effects (Sheehy *et al.*1995; Lemaire and Millard, 1999). When substrate is produced faster than it is consumed the excess accumulates as a pool of unincorporated substrate. This can be economically significant as it affects the feed value of the herbage when it is grazed or the ease with which it can be preserved as silage for Winter feeding. Both the English and Dutch models describe the changes in substrate level during growth.

All three models simulate the system at crop level rather than at the level of the individual tiller. However, in the English model the system is based on the morphology of the individual tiller so that the crop is described realistically in terms of sward structure. Each grass leaf is differentiated into an upper part which forms a flat blade and is called the lamina and a lower tubular structure called the sheath. Each successive new leaf emerges from within the tube formed by the sheaths of older leaves. This leaf morphology gives rise to the typical architecture of the grass sward where the leaf laminae form an upper layer supported on a lower layer of sheath tubes. The sheath tubes are typically 50mm in length. The separation between the two layers is relatively distinct and there is relatively little leaf lamina below 50mm. As a result of the sheath component being described explicitly, the grazing function used in the English model can be used in a realistic way to remove lamina without sheath. The production estimates given for the model are based on the estimated removals of lamina by the grazing function. The Dutch model describes leaf lamina only and at each harvest a fraction of the total of lamina is assumed to remain unharvested. The estimates of production are based on the increase in lamina mass from this post-harvest value. In the Irish model crop morphology is ignored. The estimated production for a period is the production of a morphologically undefined mass above an undefined harvest height.

These considerations form the fundamental differences between the three conceptual models of the same system. The different conceptual models lead to different mathematical models which in turn produce different outputs (but with remarkably similar trends) when compared. The following descriptions of the models should be cross referenced with the fundamentals discussed in Chapter 1 and with the meteorological data types and sources discussed in Chapter 3.

5.3   The basic conceptual framework

The most simple conceptual image of a vegetative crop is a mass bounded by two surfaces. One surface (the leaf surface) is involved in the uptake of carbon by photosynthesis and the other (the root surface) is involved in the absorption of nutrients and water from the soil. The root mass is usually a relatively small part of the system, and where there is no restriction in the supply of soil water and nutrients the root system is often ignored or treated in a trivial way (as has been done in the three models described here). The growth rate of the leaf-bearing shoot system is primarily a function of the rate of carbon capture in photosynthesis, the rate of use of the products of photosynthesis for the construction of new mass and use for the maintenance of the system. Light and temperature are primary environmental factors that affect the rates. The system rate of photosynthesis depends not only on light and temperature but also on the area of leaf surface. As growth proceeds the area of leaf and consequently the system rate of photosynthesis increases so that even in a constant environment the rate of growth in the mass of the system increases. The English and Dutch models describe the process in these terms and that is why they are described as mechanistic and dynamic. The Irish model relates growth simply to temperature and light so that it is not possible to describe growth over time. It estimates the yield existing at the end of a period subject to the average environmental conditions over the period.

5.4   The models used for comparison

5.4.1   The English model

The description of the model is based on Johnson & Thornley (1983) and on Johnson & Parsons (1985). The equations selected for presentation here are taken directly from these publications and the reader is referred to them for full details.

The first event in the model is the calculation of the amount of carbon captured by photosynthesis each day. This carbon enters a pool of substrate from which it is distributed to root and shoot. The carbon allocated to the shoot is distributed to leaf lamina production, to leaf sheath (including stem) production and to maintenance. The carbon allocated to the leaf enters the expanding leaf component and at intervals corresponding to the interval of new leaf appearance it is progressively moved between the leaf age components until it finally enters the senescent leaf compartment and is lost from the plant.

Equation 5.1 is used to calculate daily carbon capture in photosynthesis:

$$P_d = \theta \, \frac{P_m h}{k} \ln \left[\frac{\alpha \, kJ \, /h + (1 - \tau)P_m}{\alpha \, kJ \, \exp(-kL)/h + (1 - \tau)P_m} \right] \qquad 5.1$$

where
P_d = daily gross photosynthesis (kg Carbon m^2 of ground d^{-1})
P_m = maximum daily gross photosynthesis (kg Carbon m^2 of ground d^{-1})
α = initial slope of the relationship between single leaf photosynthesis and light
k = light extinction coefficient
L = leaf area index (m^2 of leaf m^{-2} of ground)
J = daily light receipt (J m^{-2} d^{-1} photosynthetically active radiation)
τ = leaf light transmission coefficient
h = day length (hours)
θ = factor to convert CO_2 to Carbon.

Equation 5.1 is the combination of several equations that deal with light interception and the relation between leaf and canopy photosynthesis. These are based on empirical relationships that have found general acceptance in crop modelling (see section 1.3.1 which describes the practical structure of a mechanistic model). It is not necessary in the present context to explain in detail how the equation is derived but it is useful to identify the functional significance of its elements.

The relation between the rate of single leaf photosynthesis and radiation incident on the leaf surface is treated as a rectangular hyperbola. The maximum rate of single leaf photosynthesis at high radiation levels is represented by P_m and α represents the initial slope of the relationship. The total radiation intercepted by the canopy depends on the radiation incident at the surface (J), on the amount of leaf surface present in the canopy (L), on the attenuation of radiation as it penetrates the canopy (k) and on the extent to which radiation is transmitted through leaves (τ). Approximately half of the radiation receipt at the crop surface is photosynthetically active radiation (PAR) and J is in terms of PAR. The equations of photosynthesis express photosynthetic rate in units of CO_2 and θ is a factor to convert CO_2 to C which is the unit used throughout the model.

In grass, a fraction of the product of photosynthesis is made available for growth and maintenance of the root. Typically the fraction has a value approximately 0.1. Apart from this allocation of substrate to the root the model deals exclusively with the above-ground shoot. In the model the remainder, $P = \phi\, P_d$ (where ϕ represents the fraction allocated to the shoot) becomes part of the pool of substrate in the shoot that is available for growth and maintenance in the shoot.

The second major event in the model is the calculation of the rate of synthesis of new structural material and the associated respiratory costs. Equation 5.2 is used to calculate the rate of synthesis of new structural material:

$$G = \mu \frac{W_s W_g}{W_s + W_g} \qquad\qquad 5.2$$

where
G = rate of synthesis of new structure (kg C m^{-2} d^{-1})
W_g = total structural weight (kg C m^{-2} d^{-1})
W_s = weight of substrate (kg C m^{-2} d^{-1})
μ = a rate constant (d^{-1}).

The equation is based on the empirical finding that the relationship between the specific growth rate of plants and substrate level in the plant may be described by this type of Michaelis-Menten equation (Johnson, Ameziane & Thornley 1983). Equation 5.2 shows that when the level of substrate in the crop is great in comparison to the structural mass, the rate of synthesis is constrained by the structural mass present. Conversely, when the substrate mass is small compared to the structural mass the rate of synthesis is controlled by the substrate level. This Michaelis-Menten equation is a way to represent the co-limitation of growth by both supply and demand. The demand function (that is the morphogenesis) is a simplified treatment of reality as it does not deal explicitly with plant form and architecture.

The synthesis of new structural material is accompanied by a respiratory cost. Typically, each unit of substrate utilised results in the production of 0.75 units of new structure. The remainder is respired. The substrate used in growth respiration is calculated as:

$$R_g = \frac{G}{Y} - G \qquad\qquad 5.3$$

where

R_g = rate of substrate use in growth respiration (kg C m^{-2} d^{-1}).
Y = the fraction of substrate in new structural material (kg C m^{-2} d^{-1}).
G/Y = the total substrate used to produce G units of structure (kg C m^{-2} d^{-1}).

The synthesis of new structural material and the simultaneous growth respiration are associated with the first leaf compartment where leaf expansion is occurring. Maintenance respiration is calculated for all four leaf compartments. Maintenance costs are always satisfied so that maintenance respiration is calculated as a function of the mass of structural material in each leaf compartment:

$$R_m = \sum_{i=1}^{4} M_i W_i \qquad\qquad 5.4$$

where
R_m = total maintenance respiration (kg C m^{-2} d^{-1})
M_i = maintenance coefficient leaf compartment i
W_i = structural weight leaf compartment i (kg C m^{-2} d^{-1}).

Due to the successive leaf compartments representing an ageing sequence the coefficient of respiration (M_i) is varied between compartments. The model also treats maintenance respiration as a process of degradation and re-synthesis of structure but the reader should refer to the original references to pursue this further.

These four equations are used to calculate the daily input of new substrate from photosynthesis to the substrate pool and the daily consumption of the pool for the synthesis of new structure for growth respiration and for maintenance respiration. The final event in the model is the adjustment of the morphology of the crop. The rate of new leaf area production is calculated as:

$$\frac{dL}{dt} = \delta \rho G \qquad\qquad 5.5$$

where
L = leaf area (m^2)
ρ = the fraction of the new shoot growth (G) that is used for leaf lamina
δ = is the ratio of leaf lamina area and leaf lamina weight (m^2 of leaf kg^{-1} C).
Senescence is represented by loss of material from the fourth leaf compartment:

$$S = \gamma W_4 \qquad\qquad 5.6$$

where
S = rate of senescence (kg C m^{-2} d^{-1})
W_4 = the weight of structure in the fourth leaf compartment (kg C m^{-2} d^{-1})
γ = a rate constant (d^{-1}).

The rate constant, γ, which is equivalent to the rate of leaf appearance, is used to control the passage of material between leaf compartments. For example, the rate of change in the mass of leaf in the third leaf compartment (kgC m^{-2} d^{-1}) is calculated as:

$$\frac{dW_3}{dt} = \gamma W_2 - \gamma W_3 \qquad\qquad 5.7$$

A value of 0.1 is typical for the parameter γ (a new leaf appears at intervals of 10 days), and the equation shows that for this value over a 10-day interval, there is complete transfer of leaf material between compartments.

Temperature affects the rate of leaf appearance (γ), the growth coefficient (μ), the maintenance coefficients (M_i) and the maximum rate of leaf photosynthesis (P_m) . It is assumed that the effect is linear as:

$$X(T) = X(20)\frac{(T - T_0)}{(20 - T_0)} \qquad\qquad 5.8$$

where
$X(T)$ = the value of the parameter at T$^{\circ}$C
$X(20)$ = the maximum value of the parameter (the value at 20°C)
T_0 = The limiting temperature for growth (0°C).

The part played by each of the equations in the overall operation of the model is evident in the following mathematical summary of the model. The rate of change in total mass is:

$$\frac{dW}{dt} = P - R_g - R_m - S \qquad\qquad 5.9$$

The rate of change in the level of substrate is:

$$\frac{dW_s}{dt} = P - \frac{G}{Y} - R_m \qquad\qquad 5.10$$

The rate of change in the weight of the first leaf compartment can be described by:

$$\frac{dW_1}{dt} = G - \lambda\gamma W_1 \qquad\qquad 5.11$$

where λ takes account of the fact that the leaves in the first compartment are expanding and the leaves to be transferred are greater than the average leaf in the compartment. The rate of change in the weight of the other leaf compartments is:

$$\frac{dW_i}{dt} = \gamma W_{i-1} - \gamma W_i \qquad\qquad 5.12$$

The leaf area associated with each compartment is treated in a similar way.

Johnson & Parsons (1985) combined the basic model with a grazing function so that it could be used to study the behaviour of grass swards in a grazing context. This was an important addition to the model because grazing management is a complex area of grassland agriculture. More parameters were introduced to cater for the distribution of the total storage between sheath and lamina and between the four leaf compartments, at different concentrations in each case. This was necessary because only leaf lamina is consumed in grazing and utilisation is not equal across leaf compartments. Grazing intake is described by an empirical equation:

$$c = c_m \frac{\left(\dfrac{L}{K}\right)^q}{1 + \left(\dfrac{L}{K}\right)^q} \qquad\qquad 5.13$$

where
c = daily intake (kg C animal^{-1} d^{-1})
c_m = maximum daily intake (kg C animal^{-1} d^{-1})
L = leaf area index (m^2 of leaf area m^2 of ground)
K = leaf area index at which intake is half of maximum
q = parameter (>1.0) that determines the rate of change in intake in response to L.

The equation describes a sigmoidal relationship between leaf area (L) and intake. At high values of L the intake c, approaches a maximum c_m. The precise shape of the curve is determined by the parameters K and q. The grazing model also introduced parameters to allow for differences in the rate of consumption of the four leaf compartments.

5.4.2   The Dutch Model

As with the English model, in the Dutch model the crop is composed of structure and a pool of substrate. Part of the substrate is allocated to root and all of the remainder to shoot. The shoot is assumed to be composed of leaf only as a single compartment. There is no distinction between lamina and sheath. The calculation of photosynthesis is based on the same parameters but the processes are not expressed as explicitly as in the English model. The calculation of growth is based on the pool size relative to demand and in a very general way is similar to the approach used in the English model. The model operates at the crop level, as in the English model, but tiller number is introduced because growth is derived from leaf extension rate per tiller. The model does not describe reproductive growth.

The description is based on Bouman, Schapendonk, Stol & van Kraalingen (1996) and the equations presented are taken directly from that publication. The symbols have been changed where appropriate to conform with those of the English model. The processes are described using units of herbage mass.

The first event in the Dutch model is the calculation of the rate of photosynthesis from radiation intercepted and its efficiency of utilisation:

$$P = E_t PAR_{int}$$ 5.14

where
P = total daily photosynthesis (g m^{-2} d^{-1})
E_t = light use efficiency in photosynthesis (g MJ^{-1})
PAR_{int} = light (photosynthetically active) intercepted (MJ m^{-2} ground).

This is based on the concept that growth is a constant proportion of the amount of radiation intercepted (Gosse et al., 1986).

The radiation intercepted is calculated as a fraction of the radiation received at the crop surface. The fraction is related to the crop leaf area and to the light extinction coefficient:

$$PAR_{int} = J(1 - e^{(-kL)})$$ 5.15

where
J = photosynthetically active radiation received at the crop surface (MJ m^{-2} ground)
k = light extinction coefficient
L = leaf area index (m^2 leaf m^{-2} ground).

According to Equation 5.15, PAR_{int} increases with radiation received at the crop surface, as the extinction coefficient increases and as the leaf area increases. Light utilisation efficiency is assumed to have a maximum value. The actual efficiency is affected by light intensity and by temperature:

$$E_t = f(T)f(J)E_{max}$$ 5.16

where
T = mean daily air temperature ($^{\circ}$C)
E_{max} = maximum efficiency of light use for dry matter production (g MJ^{-1}).

The temperature function has the value 1.0 at temperatures greater than 8°C and the value zero at temperatures less than 3°C. Between these temperatures the function is linearly related to temperature. These relations mean that photosynthesis ceases at temperatures less then 3°C and photosynthesis is independent of temperature at temperatures greater than 8°C. The light intensity function has the value 1.0 when J is less than 5.0 MJ m^{-2}. At greater values of J the function decreases linearly with J.

The second event in the model is the calculation of growth rate. The demand, that is determined by the effect of temperature on leaf growth, is compared with the supply of assimilates in the substrate pool. The use of temperature to generate the demand function represents a major difference between this model and the English model where the demand is related to plant mass. Actual growth is taken as the minimum of the demand and supply so that, as in the English model, growth is determined by the available pool when the pool is relatively small. The supply is the sum of photosynthesis and the existing pool of substrate:

$$\Delta W_s = P + \Delta W_{pool} \qquad\qquad 5.17$$

where
ΔW_s = total substrate supply (g m^{-2} d^{-1})
ΔW_{pool} = supply of substrate from pool (g m^{-2} d^{-1}).

The total demand from the pool for shoot growth and the accompanying allocation to root is calculated from the leaf extension rate:

$$\Delta W_d = \frac{1}{\phi}\left(\frac{TIL.\Delta LV.D_{lv}}{\delta}\right) \qquad\qquad 5.18$$

where
ΔW_d = assimilate demand (g m^{-2} d^{-1})
ΔLV = the leaf extension rate per tiller (cm tiller^{-1} d^{-1})
D_{lv} = leaf width (typically 3mm)
δ = the leaf area per unit weight of leaf or, the specific leaf area (m^2 leaf g^{-1})
TIL = the tiller number (m^{-2})
ϕ = the fraction of assimilates partitioned to the shoot.

The leaf extension rate per tiller varies linearly with the logarithm of temperature.

Leaf area growth is calculated from the substrate allocated to growth, ΔW_d (or by ΔW_s where growth is limited by the availability of substrate) and from the leaf death rate by:

$$\Delta L = \phi\,\Delta W_{g,s}\delta - L\left(1 - e^{(-RDR\,t)}\right) \qquad\qquad 5.19$$

where
ΔL = rate of increase in leaf area index (m^2 leaf m^{-2} ground d^{-1})
RDR = the relative rate of leaf area senescence (d^{-1})
t = time in days (d).

RDR is related to the leaf area of the crop. At L < 4, RDR takes the value 0.01, at L > 8 it takes the value 0.04 and between 4 and 8 it increases linearly with L.

In grass each newly formed leaf provides a potential site for a new tiller so that there is a relationship between leaf and tiller appearance. The extent to which sites created by new leaf appearance are filled is controlled by light competition within the crop. Tiller death through competition is also controlled by competition. In the model the change in tiller number is calculated:

$$\Delta TIL = \gamma TIL(RTR - RDR) \qquad\qquad 5.20$$

where γ is the leaf appearance rate and RTR and RDR are the relative tiller appearance and death rates respectively. The leaf appearance rate is linearly related to temperature. RTR is linearly related to L and RDR is linearly related to temperature.

5.4.3   The Irish model

The model is described in Brereton, Danielov and Scott (1996). It is a static model which was originally created as a basis for evaluating the gross effects of year-to-year differences in weather conditions on herbage production in grazing systems. It does not, and was not intended to, explain the nature of grass growth. Instead, the intention was to provide a means to understand the dynamics of a grazing management system subject to a variable feed (herbage) supply. The approach is very simple compared to the other two models but the variation between the reproductive and vegetative states of the grass crop is taken into account. Reflecting the simple empirical nature of this model the description of the system is reduced to three equations. The increase in herbage mass for a period is calculated from the radiation received at the crop surface in the period:

$$\Delta W = \varepsilon \frac{\Delta J}{Q} \qquad\qquad 5.21$$

where
ΔW = mean daily herbage dry matter yield increase (kgha^{-1})
ΔJ = mean daily light received at the crop surface (J cm^{-2})
Q = the heat of formation of plant matter (18.81 MJ kg^{-1})
ε = the efficiency of conversion to plant energy of the radiation received at the crop surface.

According to Equation 5.21 yield is proportional to the radiation received. However, it is assumed that the capacity of the crop to utilise light is saturated at a flux density of 144 J cm^{-2} hour^{-1}. In Equation 5.21 ΔJ represents light received at flux densities less than 144 J hour^{-1}. ε is not the same as E$_t$ in the Dutch model, which relates growth to the amount of radiation intercepted by leaf area. The other two models describe an effect of temperature on leaf area growth and light interception The effects of leaf area on light interception are implicit in the Irish model in the assumption that the efficiency parameter is affected by temperature. Efficiency varies with temperature in a sigmoidal relationship:

$$\varepsilon = \varepsilon_{max} \left[\frac{\left(\dfrac{T}{K}\right)^{n}}{1+\left(\dfrac{T}{K}\right)^{n}} \right]$$

5.22

where
T = mean daily temperature ($^{\circ}$C)
ε_{max} = the maximum efficiency at high temperatures
K = temperature at which ε is half maximum ($^{\circ}$C).
n = constant.

The parameters of Equation 5.22 were selected to represent a reproductive crop harvested after 28 days re-growth, fertilised with N at 600 kg N year^{-1} and adequately supplied with water and other nutrients (i.e. growth not restricted by soil nutrients or by soil water deficit). For other conditions the daily yield increase is expressed as a multiplicative function:

$$\Delta W' = \Delta W \varsigma \eta \omega v$$

5.23

where
$\Delta W'$ = daily herbage dry matter yield increase (kg ha^{-1})
ς = f(ontogeny)
η = f(nitrogen)
ω = f(soil water)
v = f(duration of growth period).

The grass crop is in the reproductive state from late Winter until June when it becomes vegetative. In the vegetative state growth rates are approximately halved compared with the reproductive crop. A sigmoid curve is used to describe the relationship between calendar date and the ontogenetic factor that represents this change:

$$\varsigma = 1 - \left[\frac{(1-\beta)\left(\dfrac{D}{F}\right)^{m}}{1+\left(\dfrac{D}{F}\right)^{m}} \right]$$

2.24

where
D = date as day number
β = the growth potential of the vegetative sward as a fraction of the reproductive sward
F = the day when the transition from reproductive to vegetative growth is half completed
m = a constant.

Nitrogen (N) fertiliser has a major effect on grassland production. To accommodate variable levels of N use the growth estimates are adjusted by η. A cubic relation between η and N is employed. The relation is based on the relationship between stocking rate and annual N requirement for grazed swards used by Teagasc.

Similarly, a multiplicative factor ω is employed to adjust yield estimates for soil water deficit. ω takes the value 1.0 where the soil water deficit is less than 30mm. At greater deficits ω decreases linearly with increasing deficit.

The logistic growth equation is used as the basis for modifying yield estimates according to the duration of the growth period. The logistic equation is commonly written in the form:

$$W = \frac{W_0 \times W_f}{W_0 + (W_f - W_0)e^{\mu t}} \qquad\qquad 5.25$$

where
W = the herbage mass (kg ha^{-1}) at t days after defoliation
W_0 = the herbage mass (kg ha^{-1}) at t =0
W_f = the ceiling herbage mass (kg ha^{-1}), approached as t becomes large
μ = the equation parameter
t = time (d).

Re-arrangement of the logistic equation for mass:

$$\mu = \frac{1}{t}\ln\left[\frac{W_0(W_f - W)}{W(W_f - W_0)}\right] \qquad\qquad 5.26$$

The 28-day yields are used to estimate μ by Equation 5.26 and Equation 5.25 is then used to estimate the yield at the relevant growth period. The seasonal change in ceiling yield is calculated using a sine-wave approximation by Equation 5.27:

$$W_f = 0.5(W_h - W_w)Sin\left[(t - 60)\frac{\pi}{182.5}\right] + 0.5(W_h + W_w) \qquad\qquad 5.27$$

where
Wf = ceiling yield (kg ha^{-1})
W_w = the mid-winter minimum ceiling mass (3500 kg ha^{-1})
W_h = the mid-summer maximum ceiling mass (7500 kg ha^{-1})
t = day of year.

5.5   Model outputs compared

The weather data for each of the five years used for the comparison are presented in Table 5.1. Variable summer rainfall was the principal difference between years. In 1993 and in 1997 there were no periods of significant drought. Significant drought occurred during the period June-August in 1994, during June-September in 1995 and during July

Table 5.1: Weather data for Grange Research Centre, 1993-1997

Mean daily temperature, ^{o}C

MONTH	1993	1994	1995	1996	1997
JAN	5.8	4.8	4.3	6.1	4.6
FEB	6.2	3.8	5.9	3.5	6.2
MAR	6.7	7.0	5.1	5.3	8.0
APR	9.1	7.0	8.5	8.4	9.2
MAY	9.7	9.3	10.6	8.4 .	10.6
JUN	13.6	12.8	13.7	13.3	12.6
JUL	14.3	15.2	16.7	15.4	15.6
AUG	13.7	14.1	17.6	14.5	16.6
SEP	11.6	11.6	12.4	13.3	13.5
OCT	7.1	9.5	12.3	10.8	10.4
NOV	5.6	9.3	7.6	5.4	8.5
DEC	4.8	6.4	4.0	4.2	6.1

Mean daily precipitation, mm

MONTH	1993	1994	1995	1996	1997
JAN	3.0	3.6	4.6	2.2	0.5
FEB	0.7	3.8	4.1	2.4	4.1
MAR	1.5	2.9	1.9	2.5	0.7
APR	2.7	2.4	1.1	2.8	1.8
MAY	3.4	1.1	1.2	2.7	2.5
JUN	4.3	1.3	0.5	0.6	3.2
JUL	2.1	2.3	2.3	2.1	1.7
AUG	1.4	1.9	0.1	3.1	2.7
SEP	2.7	2.8	2.6	0.7	1.6
OCT	3.7	1.4	3.3	3.2	2.9
NOV	1.7	1.9	6.0	3.6	3.0
DEC	4.7	3.8	2.3	1.4	3.0

Mean daily bright sunshine, hours

MONTH	1993	1994	1995	1996	1997
JAN	1.0	2.8	1.1	0.5	1.0
FEB	1.4	2.2	1.7	2.6	2.4
MAR	2.1	4.0	3.3	1.4	3.0
APR	3.5	5.6	5.0	3.6	3.3
MAY	5.3	6.1	4.7	5.6	6.4
JUN	2.9	4.2	7.2	5.9	4.0
JUL	3.1	3.5	4.4	4.6	4.8
AUG	5.7	4.3	8.1	3.1	5.0
SEP	3.1	3.5	4.1	5.0	4.4
OCT	4.2	3.8	3.2	1.8	3.2
NOV	3.1	1.5	1.9	2.4	1.4
DEC	1.1	1.6	1.4	1.0	1.3

in 1996. The estimated maximum reduction in growth rate was approximately 40% in 1994 and 1996 and more than 50% in 1995.

The measured herbage production at Grange Beef Research Centre, Dunsany, Co. Meath in the five years from 1993 to 1997, for 21-day re-growth periods and with Nitrogen fertiliser applied evenly across the year at a rate equivalent to 300 kg ha^{-1} annum^{-1} is presented in Table 5.2 and the corresponding model estimates in Table 5.3. The data of Table 5.2 represent the means of three varieties of Perennial ryegrass, the early flowering variety–Yatsin; an intermediate flowering variety–Tyrone and a late-flowering variety–Majella. In four replicate plots, 21-day re-growths for each variety were measured each week. Both measured and calculated herbage yields are presented as the means of three consecutive weeks. In all years the grasses were in the first year of growth after sowing. Sowing was carried out in the autumn preceding the year of measurement.

In the case of the English model the data presented represent the removals by the grazing function at 21-day intervals. The grazing function was switched off when the leaf area index was reduced to 1.2. In the case of

Table 5.2: Herbage mass and harvest dates for grass at Grange Research Centre, 1993-1997

Herbage mass, dry matter, kg ha^{-1}

1993	1994	1995	1996	1997
1301	77	26	50	36
2076	164	219	135	490
2120	1044	766	1288	1278
1969	1950	1139	1604	1613
1671	1085	1320	2043	1556
1408	1747	1620	1326	1807
1112	1717	784	1155	1442
642	1380	1204	1245	1791
267	880	1011	1419	1415
	976	136	1431	995
	477	468	607	954
		754	366	473

Annual total

12566	11498	9445	12670	13848

Harvest dates

1993	1994	1995	1996	1997
25 Apr	14 Mar	28 Feb	10 Mar	7 Mar
16 May	4 Apr	21 Mar	31 Mar	28 Mar
6 Jun	25 Apr	11 Apr	21 Apr	18 Apr
27 Jun	16 May	2 May	12 May	9 May
18 Jul	6 Jun	23 May	2 Jun	30 May
8 Aug	27 Jun	13 Jun	23 Jun	20 Jun
29 Aug	18 Jul	4 Jul	14 Jul	11 Jul
19 Sep	8 Aug	25 Jul	4 Aug	1 Aug
10 Oct	29 Aug	15 Aug	25 Aug	22 Aug
	19 Sep	5 Sep	15 Sep	12 Sep
	10 Oct	26 Sep	6 Oct	3 Oct
		17 Oct	27 Oct	24 Oct

the Dutch model the data presented represent the removal of leaf area to a residual leaf area index of 1.0 at 21-day intervals. In both the English and Dutch models the part of the pool of substrate that was in the lamina was included in the value given for 21-day yield. In the Irish model the data represent the accumulated yield over 21 days. Yield was estimated daily. The logistic function was used to adjust the daily estimates for the number of days of re-growth.

Each of the three models was programmed using parameter values given in the source publications. The weather data used (Table 5.1) were daily values from a climate station located at the research centre where the herbage yields were measured. Temperature values were the mean of the maximum and minimum air temperature and sun hours were measured using a Campbell-Stokes glass sphere. Daily radiation receipt was calculated from sun hours using the Angstrom formula (Martinez-Lozano, Tena, Onrubia & De la Rubia 1984) with the parameters proposed by McEntee (1980). For all years, the relative reduction in growth rate during drought estimated by the Irish model was applied to the data from the other models.

The effect of these droughts is reflected in the annual total herbage harvested where the lowest totals, in 1994 and 1995 (Table 5.2), were associated with the years of most prolonged and most severe drought. The variation in model estimates of total yield between years was generally correlated with measured production. In all cases the greatest yield was in 1997 and the lowest in 1995. The English model consistently overestimated total annual yield by approximately 50%. The overestimate of annual

Table 5.3: Herbage mass calculated by the three models

English model

1993	1994	1995	1996	1997
5244	3104	604	1242	1468
3020	1627	1343	1216	1175
1894	3229	1845	2111	1670
1268	2987	2575	2703	2091
1766	2724	2307	3554	3256
1471	1274	1455	2401	2312
2051	872	1518	991	1745
732	1122	772	954	1434
747	1237	1470	929	1099
	1208	312	1323	1471
	554	460	673	702
		407	49	175

Annual total				
18193.26	19938.06	15066.6	18145.88	18598.87

Dutch model

1993	1994	1995	1996	1997
1242	372	38	52	294
1642	644	169	195	775
1672	1031	863	945	1105
1846	1738	1273	1138	1391
1926	1473	1207	1708	1803
1763	1205	1269	1786	1756
1688	1269	980	1084	1793
1034	1400	1119	1241	1642
931	1274	1058	1461	1459
272	1240	435	1393	1484
	815	658	899	1139
	485	758	383	521

Annual total				
14017	12946	9827	12285	15162

Irish model

1993	1994	1995	1996	1997
1100	496	196	212	442
1442	669	334	337	804
1369	879	928	915	990
1417	1480	1144	968	1221
1351	1153	1018	1439	1643
1119	955	1042	1553	1461
987	903	819	801	1347
539	880	772	805	1101
467	747	730	895	907
196	668	254	784	845
109	443	363	477	632
	303	447	279	338

Annual total				
9792	9274	8047	9466	11731

herbage production by the English model was almost entirely due to overestimates of the 21-day re-growths in the period of reproductive growth up to mid-June in all years (Tables 5.2 and 5.3). The model was assembled specifically to describe the growth of grass in the vegetative state and the agreement between the model and measured production in the period of vegetative growth (from mid-June) was very close. The estimates from the Dutch model were close to the measured yield in all years The agreement between the model estimates and measured annual production reflected good agreement throughout the season. The Irish model consistently underestimated yield by about 15%. The underestimate occurred uniformly across the season.

The general pattern of the seasonal distribution of growth is illustrated by 5-year averages in Figure 5.1. Data points represent the averages of all 21-day herbage yields occurring in successive half-month periods. The two vertical bars in Figure 5.1 represent the division of the season into three periods corresponding to the three distinct states of the herbage. In the initial period to mid-April the herbage is in the reproductive state. The formation of the reproductive morphology (stem extension and formation of the seed-head) has not begun and the herbage remains short and leafy. In the second period, between mid-April and mid-June, the morphology of the sward changes with the formation of reproductive features. After mid-June the herbage is returned to the short leafy morphology characteristic of the vegetative state.

The great overestimate of reproductive growth (Figure 5.1a) by the English model, a vegetative model, was not expected. The seasonal rise and fall of

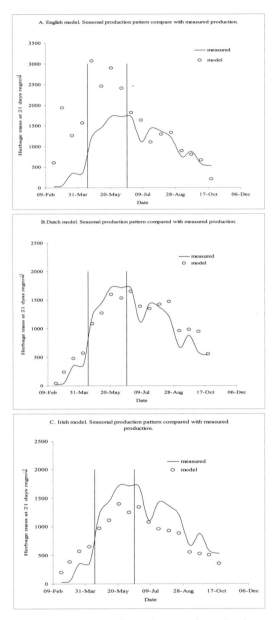

Figure 5.1: Comparison of seasonal production pattern by model and measured. A: English model. B: Dutch model. C: Irish model

air temperature lags behind the rise and fall of radiation receipt so that the ratio of radiation and temperature is greater in the first half of the season than in the second half. It has been proposed by Parsons & Robson (1982) that the greater growth potential of grass in Spring and early-Summer is partly attributable to this difference in the balance between radiation and temperature. In Spring and early-summer the relatively low temperatures lead to a relatively reduced respiratory demand. Successive leaves produced in Spring and early-Summer increase in size so that the size of the oldest leaf compartment becomes relatively smaller and the canopy is dominated by younger leaves. The maintenance requirement of the younger leaves is less. As a result of the increased proportion of older leaves, tissue death lags behind tissue synthesis.

Although the English model describes no reproductive features, the unexpected tendency of the model to overestimate rather than underestimate production in Spring may be explained by these mechanisms. It is possible that the difference between measured herbage and model estimate represents the "cost" to the grass plant of the formation of a reproductive apparatus. The English model has been used in the analysis of various aspects of the behaviour of grass swards (Thornley 1998; Parsons, Johnson & Harvey 1988; Thornley & Cannell 1997) where accurate prediction is less relevant than the establishment of understanding. This suggestion serves to illustrate the great potential of modelling to increase our understanding of the nature of biological systems.

The Dutch model exhibited the expected trend of an underestimate in the second period when reproductive structures were being formed and an overestimate in the following vegetative phase (Figure 5.1b). The Dutch model is currently used by the MARS project to monitor herbage production across Europe (Vossen & Rijks 1995). The annual production estimates of the model are in reasonable agreement with measured but the seasonal bias is evident.

The Irish model overestimated in the first period and then underestimated for the remainder of the year (Figure 5.1c). The general underestimate of production by the model may reflect that the model was parameterised in the context of old established Perennial ryegrass pastures which are less productive than newly re-seeded pasture. This suggests that the model provides a reasonable measure of the seasonal pattern of herbage production. The Irish model has been used to evaluate the scale of regional and inter-annual variation in annual grass production in Ireland. The calculated yields agreed with field measurement (Brereton 1995).

5.6   Conclusions

The conclusions that can be drawn from these comparative illustrations of agro-meteorological modelling are: (i) a single system can generate multiple conceptual models, each of which can be "correct"; (ii) the conceptual model will dictate the type of mathematical formulation that is possible; (iii) complexity will not necessarily yield a better result, or more usable model; (iv) good quality data are needed to ensure that the model has the "best chance"; and (v) when interpreting the output of the model it is necessary to fully understand the inputs and assumptions.

The example outlined represented an almost ideal situation where data were available for both the product (grass) and the driving variables (weather). The reader should consider how much more difficult such a comparative exercise would be without such a close geographical link. In attempting to apply the types of model presented here the user should pay close attention to the availability of data and how it can be interpolated (see Chapters 1, 2, 3 and 4).

References

Bouman, B.A.M., A.H.C.M. Schapendonk, W. Stol and D.W.G. van Kraalingen. (1996): *Description of the growth model LINGRA as implemented in CGMS.* Quantitative Approaches in Systems analysis No.7. DLO Research Institute for Agrobiology and Soil Fertility. Wageningen.

Brereton, A.J. (1995): Regional and year-to-year variation in production. In: D.W. Jeffrey, M.B. Jones and J.H. McAdam (Editors): *Irish Grasslands, Their Biology and Management*. Royal Irish Academy. 12-22. Dublin.

Brereton, A.J., S.A. Danielov and D. Scott. (1996): *Agrometeorology of Grass and Grasslands for Middle Latitudes*. Technical Note No.197. World Meteorological Organisation. Geneva.

Gosse,G., C. Varlet-Grancher, R. Bonhomme, M. Chartier, J-M. Allirand and G. Lemaire. (1986): Production maximale de matiere seche et rayonnement solaire intercepte par un couvert vegetal. *Agronomie* **6**: 47-56.

Johnson, I.R and A.J. Parsons. (1985): A theoretical analysis of grass growth under grazing. *Journal of Theoretical Biology* **112**: 345-367.

Johnson, I.R. and J.H.M. Thornley. (1983): Vegetative crop growth model incorporating leaf area expansion and senescence, and applied to grass. *Plant, Cell and Environment* **6**: 721-729.

Johnson, I.R., T.E. Ameziane and J.H.M. Thornley. (1983): A model of grass growth. *Annals of botany* **51**: 599-609.

Lemaire, G. and P.Millard. (1999): An ecophysiological approach to modelling resource fluxes in competing plants. *Journal of Experimental Botany* **50**: 15-28.

Martinez-Lozano, J.A., F. Tena, J.E. Onrubia and J. De la Rubia. (1984): The historical evolution of the Angstrom formula and its modifications; review and bibliography. *Agricultural and Forest Meteorology* **33**: 109-128.

McEntee, M.A. (1980): A revision of the equation relating sunshine hours to radiation income for Ireland. *Irish Journal of agricultural Research* **15**: 223-236.

Parsons, A.J. and M.J. Robson. (1982): Seasonal changes in the physiology of S24 Perennial ryegrass (Lolium perenne L.) 4. Comparison of the Carbon balance of the reproductive crop in Spring and the vegetative crop in Autumn. *Annals of Botany* **50**: 167-177.

Parsons, A.J., I.R. Johnson and A. Harvey. (1988): Use of a model to optimise the interaction between the frequency and severity of intermittent defoliation and to provide a fundamental comparison of the continuous and intermittent defoliation of grass. *Grass and Forage Science* **43**: 49-59.

Sheehy, J.E. , P.L. Mitchell, J-L Durand, F. Gastal and F.I. Woodward. (1995): Calculation of translocation coefficients from phloem anatomy for use in crop models. *Annals of Botany* **76**: 263-269.

Stapleton, J. and M.B. Jones. (1989): Effects of vernalisation on the subsequent rates of leaf extension and photosynthesis of perennial ryegrass (Lolium perenne L.). *Grass and Forage Science* **42**: 27-31.

Thornley, J.H.M. (1998): *Grassland dynamics, an ecosystem simulation model*. CAB International. Wallingford.

Thornley, J.H.M. and M.G.R. Cannell. (1997): Temperate grassland responses to climate change: An anlysis using the Hurley pasture model. *Annals of Botany* **80**: 205-221.

Vossen, P. and D. Rijks. (1995): *Early crop yield assessment of the EU countries: The system implemented by the Joint Research Centre*. Publication EUR 16318 of the Office of the Official Publications of the EC. Luxembourg.

Chapter 6   A   GIS   based   Model   of   Catchment Surface Water Quality[†]

M. Bruen, A. Dowley and J. Zhang
Department of Civil Engineering, University College Dublin, Ireland.

6.1   Introduction

This chapter describes one of the steps involved in the development of a Decision Support System (DSS) for managing river water quality based on a Geographic Information System (GIS) (see Chapter 2 for background information). The final package is intended for use in practice for catchment management, for development planning and as a research tool. The DSS is to be developed in three phases. Only the first phase is described here which required building a GIS based tool with analytical capabilities. It integrates hydrological and hydraulic models, non-point source pollution models and river water quality models. This first phase provides enough information to help a decision maker to make reasonable and efficient decisions about water management at a catchment scale. Phase two will require the implementation of design or synthesis capabilities into the system and phase three will add multi-criteria decision support tools into the package.

In general, the water management of a catchment must addresses issues relating to the quantity and quality of both surface waters and groundwater and of interactions between them. In the work described here, we concentrate on surface waters. Figure 6.1 illustrates management tasks with respect to surface water, i.e. surface flow and its quality which are important for catchment management in agricultural areas. The two important time scales are:

(i)   *Long-Term Management*: involving yearly or monthly water supply plans, optimal operation of reservoirs, water resources assessment and drought analysis. All these require yearly, monthly or longer period (e.g. the flood period) flow forecasting.

(ii)   *Short-Term Management* for which the most important task is flood forecasting and control.

Water pollution in a catchment can be categorised as: (1) Point-source pollution, i.e. continuous discharge of pollutant at one point, for example, an industrial effluent or (2) Non-point source or diffuse pollution, i.e. the pollution caused by agricultural activity, the population of humans and animals, soil erosion over the catchment or atmospheric inputs. Models are required for all these elements and the GIS is used as the central link which integrates all the individual models and controls their activities.

[†] Some of the work described here was supported by Dublin Corporation Main Drainage Division and the European Union Copernicus programme

Meteorological inputs drive this system: precipitation and potential evapotranspiration are the most important dynamic data requirements for any model of a catchment. The scale and amount of data required depends on the purpose of the analysis, type of model used and on the size of the catchment. In general, the shorter the response time of the catchment, the smaller the time-step required both for model calculations and input data. For many small catchments in Ireland the amount of direct hydrological and meteorological data available is less than ideal and special techniques for record extension or data transfer may be required.

This chapter begins by reviewing GIS and decision support in the hydrological sciences, then outlines each of the important hydrological models used in the Decision Support System and finally describes its application to the Dodder catchment in eastern Ireland.

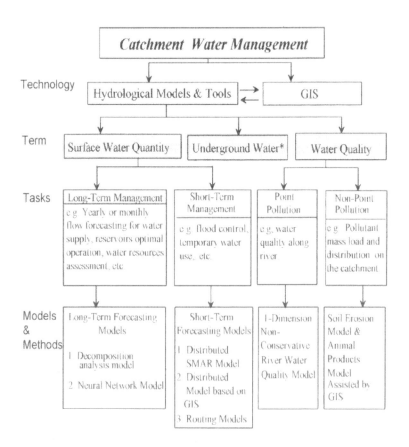

Figure 6.1: Catchment water Management Tasks

6.2   Review of GIS Applications in Hydrology

In the early 1980s GIS was introduced as a means of managing and presenting spatial data. Its use has become widespread for management, processing, simulation, and visualisation of spatial data in hydrological science and in water-related disciplines. Hydrology is the science of the waters of the Earth, their occurrence, circulation and distribution, their chemical and physical properties, and their reaction with their environment. As an Earth science, there is a close relationship between hydrology and geography. The capabilities of GIS with regard to input, management, analysis and presentation of the spatial geo-referenced data of a catchment, has encouraged hydrologists to conduct theoretical research into basic hydrological processes.

Applications have been developed in many areas. Examples of using GIS as a management tool include:

- Assessment of groundwater contamination potential (Bruen & Masopha, 1991; Engle *et al.*, 1996);
- Surface flow modelling, (TOPMODEL - Romanowicz *et al.*, 1993; SHE model - Abbott *et a.l*, 1986(a)(b); MIKE-11 - Paudyal *et al.*, 1995).
- Groundwater flow modelling (Lieste, 1993; Nachtnebel, 1993a); Kamps *et al.*, 1996; Michl, 1996).
- Water quality modelling (De Roo, 1996; Engel *et al.*, 1993; De Roo, 1993, Heidtke *et al.*, 1993; Gupta, 1977).
- Water Resources Planning, Operation and Management - regional scale (Keser, 1993; Kaden, 1993), basin scale (Lange & van der Meij, 1993; Olivier and McPherson, 1993; Costa *et al.*, 1996) and expert systems (Lam, 1993; Furst, 1993; Frysinger, 1993; Jamieson, 1996; Fedra & Jamieson, 1996b).
- Study of spatially distributed unit hydrographs (Maidment, 1993).
- Modelling of the spatial variability of hydrological processes (Moore *et al.*, 1993).
- Forest effects on hydrology (van der Sel *et al.*, 1993).
- Global climate modelling (Skelly *et al.*, 1993).

6.3   Decision Support Systems and their Applications in Hydrology and Water Resources

A decision support system can be defined as `*An interactive computer-based tool, which helps the decision maker utilise data and models to solve unstructured problems.*' (Turban 1993). It is an interactive, flexible, and adaptable information system, specially developed for the solution of particular management problems. Taken to its limit, a DSS can be regarded as a form of artificial intelligence in which computers are used not only to predict likely consequences of decisions but also to supplement management experience in decision-making (Jamieson, 1996a). The basic unit of a DSS is the model. A complex DSS will contain a number of models integrated to describe a system, and which can be used to assist in making decisions about the system. A DSS has the following characteristics (Somlyody & Varis, 1992): (i) the ability to addresses complex management problems; (ii) the ability to couple models and other analytical tools for information management; (iii) easy operation and interaction; (iv) flexibility and adaptability. The subsystems of a DSS are: (1) *data management*, which includes maintaining the database(s) and arranging for queries and data input/output. In this study, the database management system was INFO which is part of the ARC/INFO package; (2) *model management*, which is the software package that implements the quantitative models and analysis tools and arranges for the required data inputs (from the data management subsystem) and processes the outputs; (3) *user interface*, which is also called the communication subsystem. The user can communicate with and command the DSS through this subsystem. Here the ARC/INFO macro language AML was used to develop the user interface; (4) *the report generator* which post-processes and presents data and model output in useful graphic and/or tabular forms as required by the user; and (5) *knowledge management*, which is an optional subsystem included in some complex DSSs. It contains knowledge or experience which supports the decision making. It can support any of the other subsystems or act as an independent component. In agricultural modelling this is a frequently overlooked but potentially powerful capability.

Decision-making for water management in a catchment is a complicated process. The complexity of the issues to be resolved makes it difficult to reach rational and effective decisions, and quite different decisions may be made by different decision makers. Therefore, a DSS should be developed with a number of objectives in mind. The provision of comprehensive, objective, and reliable information with the aim of improving the quality of decision-making is a priority – the system should focus the mind on that which is important. All the necessary models and analysis tools should be integrated into a single package to make using it a non-difficult task. All of these models and tools can be executed through a user-friendly interface which makes the system easy and flexible to operate, without in depth modelling knowledge, such that the same results for a particular problem are obtained by different users. Although the operation of a DSS does not require any special modelling skills on the part of the user, the system should encourage the user to incorporate prior modelling and management experience into the decision making. When developing a `full-facilities' DSS for general water management, different scales of analysis, data and models are required for different decision situations for the same catchment. Therefore, the system should be able to consider most situations and to meet those requirements usually encountered in the decision-making procedure for a general water management problem. The system architecture of the system for which some of the models described here were integrated (DSS-WMC - Decision Support System for Water Management at Catchment scales) is illustrated in Figure 6.2

6.4   The catchment models used to predict surface water quality

The catchment was described by a number of models as indicated by Figure 6.2 (the outline of the overall conceptual model (see Chapter 1). Some of the more important models are described below along with an evaluation of why they were the most suitable models, what difficulties were encountered when implementing them, and their role in the DSS. The DSS has models which can be used for gauged catchments, for which there are measured data for calibrating the model parameters. However it also has models which can be used for catchments for which there are no data. The SMAR model, described in 6.4.1 below, is an example of the former while the SCS method, described in section 6.4.3, is an example of the latter.

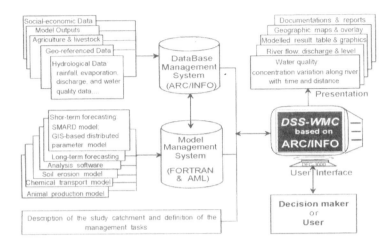

Figure 6.2: Basic Architecture of the DSS-WMC

6.4.1   The SMAR Model

The SMAR (Soil Moisture Accounting and Routing) model, also known as the Layers model, is a fairly simple conceptual model. It was originally formulated by O'Connell *et al.* (1970), and was further developed in the Department of Engineering Hydrology, University College Galway (O'Connor, 1992). It has been tested with data from catchments of different sizes and from different climatic regions (Khan, 1989; Kachroo, 1992; Zhang, 1994; Liang, 1995). There are a vast number of different models available for predicting the fate of water in catchments. Some of those with European origins include HYRROM (Institute of Hydrology), NAM now MIKE11 (Danish Hydraulic Institute), HBV (Swedish Meteorological and Hydrological Institute), ARNO (Univ. of Bologna). SMAR was chosen because of its simple structure and because two of the present authors were involved in its development.

The SMAR model is a lumped rainfall/evaporation/discharge model which means that the rainfall information for all the gauges in and around the catchment are combined for each time step to make a single "lumped" rainfall series for input to the model. Similarly a single potential evapotranspiration input series is produced from the measured data. Such a model does not have any information about the spatial distribution of any of its inputs or parameters. The SMAR model transforms the input variables, i.e. the time series of rainfall and evaporation into a discharge time series through the following steps. (Note the similarities with the model example presented in Chapter 1).

In the SMAR model, the potential evaporation depth, E_p, at a given time interval, is taken as the measured Pan evaporation depth, E, multiplied by a conversion parameter, T, i.e. $E_p = E*T$.

In determining the corresponding actual evaporation depths E_c, it is assumed that the catchment is analogous to a vertical stack of horizontal soil layers, of total moisture capacity Z (mm). Each layer of soil is assumed capable of containing up to a certain amount of water at field capacity, e.g. 25 mm in this study. Any evaporation from the top layer is assumed to occur at the potential rate, and from the second layer (only on exhaustion of the storage in the first layer) at the potential rate multiplied by a model parameter, C, whose value is less than unity. On exhaustion of the storage in the second layer, any evaporation from the third layer is assumed to occur at the potential rate multiplied by C_2 and so on. Thus, a constant potential evaporation applied to the basin would reduce the soil moisture storage in a roughly exponential manner, and the evaporation ceases when the total storage in all the layers is exhausted. When the rainfall rate, R, exceeds the potential evaporation rate E_p, a fraction, H', of the excess:

$$X = (R - E_p)$$

6.1

which is taken as being proportional to the ratio of the available water content in the top five layers to their total capacity depth of 125 mm, becomes the first component of generated runoff, denoted, i.e.:

$$r_1 = H'X$$

6.2

and H' is calculated from the equation:

$$H' = H \frac{\sum_{i=1}^{5} S_0(i)}{125}$$

6.3

H is a model parameter to be optimised. Of the remaining proportion of rainfall excess, $(1 - H').X$, anything in excess of the infiltration rate, Y, also contributes a component of generated runoff, denoted by r_2. It is expressed as:

$$r_2 = (1 - H')X - Y$$

6.4

while the remainder restores each layer to its field capacity from the top downwards until all the rainfall is exhausted or until all the layers are at field capacity. Any still remaining surplus of rainfall excess, beyond that required to fill all the layers, contributes a third generated runoff component r_3, which is moisture in excess of soil moisture capacity. A fraction, G, of this goes into groundwater, i.e.:

$$r_g = r_3 G$$

6.5

The remaining fraction:

$$r'_s = r_3 (1 - G)$$ 6.6

is added to r_1 and r_2 to give an overall surface runoff generated component:

$$r_s = r_1 + r_2 + r'_s$$ 6.7

Thus, the total generated runoff produced by the water balance component of the SMAR model is:

$$r = r_s + r_g = r_1 + r_2 + r_3$$ 6.8

The routing effects (i.e. the diffusion and/or attenuation) of the catchment on the two overall components of generated runoff (r_s and r_g) are assumed to be conservative, linear and time-invariant. The surface runoff r_s is transformed into a surface discharge component using the Nash `cascade of equal linear reservoirs' model (Nash, 1960). The outputs of the routing elements are finally added together to give the estimated (i.e. the simulated) outflows of the SMAR model. A schematic diagram of the SMAR model is shown in Figure 6.3.

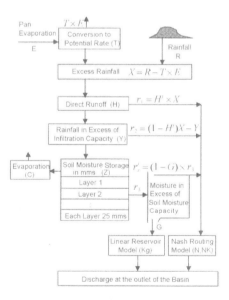

Figure 6.3: A schematic representation of the SMAR model

6.4.2 Distributed Model Based on the SMAR Model

Being a lumped daily model, it is difficult to apply the SMAR directly to a large catchment where many of the important hydrological variables, processes and inputs vary in space as well as with time. To eliminate the effect of this kind of non-homogeneity, the original lumped SMAR model was revised into a distributed model, denoted as SMARD. This modification makes it possible to apply SMAR to a large non-homogeneous catchment. In addition, the original daily model was revised into a flexible time interval model.

To apply the SMARD model, the study catchment is divided into a number of sub-basins and a separate SMAR model is fitted to each. A channel flow routing model routes the discharge from each sub-basin to the outlet of the catchment. The channel model chosen for linking the distributed SMAR models was the Muskingum-Cunge routing method, which can be used for channels where the hydrological data are sparse but basic hydraulic data about the channel are available (Cunge, 1969). This is a hydraulic routing method where flow is calculated as a function of time at cross-sections along the channel.

6.4.3   The Runoff Generating Model-SCS Runoff Curve Number Method

Catchment models, such as the SMAR described above, have parameters which must be calibrated for every catchment in which the model is used. Where there are insufficient data, or perhaps none, i.e. an ungauged catchment, available for calibrating a catchment model, some other technique must be used in the DSS to estimate runoff from rainfall. One such technique is contained in the Flood Studies Report (NERC 1974). Another is the American SCS (Soil Conservation Service) runoff curve number technique. Both have been implemented in the DSS, but for brevity only the SCS method is described here. It can be expressed as (Rawls *et al*, 1992):

$$R = \frac{\{P - I_a\}^2}{\{P - I_a + S\}}$$
6.9

where, R (mm) is the total direct runoff, P (mm) is the total storm precipitation, I_a (mm) is the initial abstraction and S (mm) is the soil moisture storage capacity. The initial abstraction I_a is a function of soil type and structure and antecedent soil moisture, and no runoff occurs until rainfall equals the initial abstraction. The soil moisture storage capacity, i.e. the potential maximum retention, S (in mm), can be defined by:

$$S = \frac{25400}{CN} - 254$$
6.10

where CN is the SCS dimensionless runoff curve number, which is defined as a function of soil hydrology group (SHG), land use (LU) (Rawls,1992), and antecedent moisture condition (AMC) (Chow, 1964).

For application in eastern Ireland, the CN in each sub-basin was determined from the land-use and soil association, after Rawls (1992). The mean CN for a sub-basin was computed using the GIS (see Chapter 2 for information on basic GIS operations and data types) and the following steps:

1.   Obtain a digital map of land-use and soil type
2.   Create a new map of sub-basins, each with land-use and soils defined
3.   Define areas within the sub-basins which each have a single land-use and soil type
4.   Determine the values of CN for each soil code and land-use code according to the Handbook of Hydrology (Maidment, 1992). This is the standard technique used in the US for ungauged catchments.

5. The average value of CN in a sub-basin is calculated within the GIS using an aerial weighted algorithm whereby each homogeneous area within the sub-basin has its contribution weighted by its fraction of the total area (the same principle as distance weighted interpolation).

Once the curve number is obtained total runoff volume for any given rainfall event for each sub-basin can be predicted. The runoff generated in each sub-basin must be routed to the basin outlet. This is done with a unit hydrograph, which is defined as the runoff hydrograph from a single rainfall event of unit depth and specified duration. For instance a one-hour unit hydrograph is the runoff which would be produced by unit depth of effective rainfall occurring uniformly over one hour. It is a very useful concept if the catchment behaves linearly, since the runoff can be obtained by multiplying the unit hydrograph by the actual depth of rain. Furthermore, the runoff from a complex rainstorm of many hours duration can be obtained by adding together the runoff produced by each individual one-hour of rainfall volume. There are two methods of generating a unit hydrograph for the ungauged sub-basins, either the unit hydrograph (UH) of the Natural Environmental Research Council of the UK, (FSR-UH), or the unit hydrograph of the Soil Conservation Service of the USA, (SCS-UH). Both of these methods only require the determination of a basin time-to-peak, T_p parameter, which are simple triangular approximations. They are only unit hydrographs for routing of direct runoff. However, the runoff generated using the SCS runoff curve number method is a total amount. It includes direct runoff, runoff drained from the unsaturated zone, called subsurface runoff, and groundwater runoff. In order to take account of the different concentrating velocities of different runoff components, a UH was proposed with two different slopes in the declining portion after the peak. The justification for this is that it is assumed that the routing of direct runoff stops at the point T_p+T_1 after which time only subsurface runoff routing or groundwater runoff routing occurs (with a UH coordinate value q_1). The parameters q_1, T_1, and T_2 reflect the contribution of sub-surface runoff and groundwater runoff to the total generated runoff and their routing velocity. They are functions of soil type, land use and geological features in the catchment. As described above, CN is a synthetic parameter reflecting the soil type and land use. In practice, CN values have a range of 30-100 (Rawls, 1992), and for the same land use, the higher the infiltration rate, the greater the groundwater runoff potential, and the smaller the value of CN. From the cumulative generated runoff volume, we obtain the runoff for each time step which is then routed to the outlet of the catchment using the Muskingum-Cunge hydraulic routing methods.

6.4.5   Non-point Source Pollution Prediction Models – Surface Water Quality

Once a satisfactory runoff and flow modelling capability has been developed, water quality issues can be addressed. Pollution from land management activities can have a major impact on water quality in a catchment. Very often a distinction is made between point sources of pollution, (septic tank, slurry pit) and diffuse sources (spreading of slurry, pesticides). Identification of the problem areas and the management of non-point sources of pollution are technically complex tasks, because pollutant sources often are located over a large geographical area and are not readily identifiable. To identify potential problem areas within a catchment, models of soil erosion and chemical movement can be used. The use of these models, has often been limited as a result of their large data requirements and the difficulty of parameter estimation. By using a GIS, effective assistance in parameter estimation can be obtained. In order to develop a DSS for catchment water quality, a number of non-point source pollution models needed to be integrated. Factors influencing non-point source surface water pollution include available water for transport, soil erodibility, and nutrients and organic material availability. These can be predicted from water availability, soil type, land use vegetation cover and livestock rates (Figure 6.4).

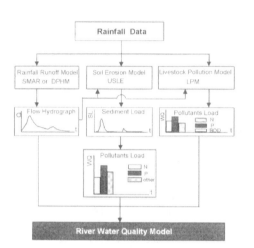

Figure 6.4: Integration of models to predict surface water quality

Soil Erosion by Water
Water is the most widespread agent of erosion. Erosion by water can be classified into two types: sheet erosion and rill erosion. Sheet erosion is the detachment of land surface materials by raindrop impact and its subsequent removal by overland flow. The transport capacity of thin overland flow, usually called sheet flow, increases with field slope and flow discharge per unit width. Rill erosion is the removal of soil by concentrated sheet flow.

The splash erosion rate depends on the raindrop velocity and diameter, the rainfall intensity, the surface cover and soil type, and can be derived from meteorological, soil and land use and management data. Runoff erosion depends on the soil type and flow-velocity. The latter is a function of land slope which is estimated in the GIS from the its digital elevation map of the catchment. The final sediment yield is however limited by the sediment transport capacity of runoff which depends mainly on the overland flow velocity which is calculated by the model in the course of its simulation.

There are many methods for estimating the soil erosion amounts and non-point source pollution loads. These methods can be classified into two categories: empirical statistical methods (which have the advantage of fitting well in situations similar to those used to derived the equations, but which can be unreliable in other situations) and physical mathematical (mechanistic) models or process-based models (which have the advantage of greater general applicability but the disadvantage of requiring considerably more data)(see Chapter 1 for more detail of model types). The method chosen was the empirical Universal Soil Loss Equation (USLE), described below, because of its simplicity and modest data requirements.

The Universal Soil Loss Equation (Wishmeier and Smith, 1978) is one of the most widely used methods of estimating soil erosion. It computes the soil losses for an individual storm event as the product of five factors: rainfall erosivity R_s, soil erodibility K, field slope-length L_s, crop management factor C, and conservation practice factor U:

$$SL = R_s K L_s C U \qquad 6.11$$

where SL is the soil loss per unit area for a single storm. Specific equations are prescribed for estimating rainfall erosivity from rainfall intensity and for estimating the slope length factor. Crop management and conservation practice factors can be estimated from tables in Shen & Julien (1992).

Agricultural Non-point Source (AGNPS) model – chemical transport in runoff
Nutrients are chemicals essential for plant growth. Two problems are associated with excessive levels of nutrients in the water environment; (1) Water may be toxic when the concentration of certain nutrient forms exceeds a critical level, and (2) eutrophication, the onset of undesirable processes in surface waters due to excess nutrient supply. The nutrients considered in this study include nitrogen (N) and phosphorus (P), both essential plant nutrients and major contributors to surface water pollution.

Sediment can be the major transporting vehicle for phosphorus and organic nitrogen. Raindrop splash and flowing water detach soil particles and organic matter containing nitrogen and phosphorus. The transport capacity of the flowing water depends primarily on the discharge and the velocity of the flow. Whenever the velocity is reduced, such as by a flatter slope, the transport capacity is reduced and any sediment in excess of the reduced capacity settles out.

Natural soils typically contain 0.05 to 0.3 percent nitrogen and 0.01 to 0.13 percent phosphorus (Knisel 1980). The application of fertilisers, manure, wastes, and crop residues increases the N and P content above natural levels while intensive cropping without nutrient additions reduces the N and P content. Direct measurement of the N and P contents in the specific fields modelled is highly desirable.

The enrichment of sediment occurs due to selective erosion and deposition processes. In the deposition process, the larger and heavier particles settle out first, the remaining sediment contains a larger percentage of finer particles which have a higher capacity per unit of sediment to absorb phosphate and organic nitrogen. Thus the transported sediment is richer in phosphorus and nitrogen than the original soil. It is suspected that changing soils, crops, or management practices should result in different amounts of enrichment.

The AGNPS model (Young, *et al*, 1989) uses the SCS runoff curve number method to estimate runoff and the USLE model to estimate the soil erosion. Chemical transport calculations are divided into soluble and sediment absorbed phases. Nutrient transported in the sediment absorbed phase is calculated using total sediment moved in a sub-basin:

$$Nut_{sed} = Nut_f \, Q_s(x) E_R \qquad 6.12$$

where Nut_{sed} (in ppm) can be nitrogen (N) or phosphorus (P) transported by sediment; Nut_f is N or P content in the field soil; $Q_s(x)$ is sediment yield calculated using the USLE model The transported sediment may not have the same concentration of nutrient (N or P) as the parent soil and the relationship is expressed as the enrichment ratio, E_R, and is calculated by:

$$E_R = 7.4 Q_s(x)^{-0.2} T_f \qquad 6.13$$

where, T_f is a correction factor for soil texture. Soluble nutrient contained in runoff is estimated from:

$$Nut_{sol} = C_{nut} Nut_{ext} Q \qquad 6.14$$

where Nut_{sol} (in ppm) can be the concentration of soluble N or P in the runoff; C_{nut} is the mean concentration of soluble N or P at the soil surface during runoff; Nut_{ext} is an extraction coefficient of N and P for movement into runoff; and Q is the total runoff. The procedures for estimating values for these parameters are given in Young *et al.* (1989)

Livestock Pollution Model
Livestock waste is another source of pollution. A livestock pollution evaluation model was included in the DSS. It consists of two parts, a short screening form and an evaluation procedure. The screening form includes: (1) the area of the sub-basin, assuming that animals are even distributed over the sub-basin; (2) the number of animals; and (3) the types of animal.

Surface flow is the major transport vehicle for animal pollutants. The amount of pollutant load carried to a channel is proportional to the depth of runoff. In the evaluation procedure, if there is no runoff (output of rainfall-runoff model), then no pollutant flows to the channel, but the pollutants are accumulated in the sub-basin, i.e. $PAL(i) = 0$, and:

$$SAL(i) = SAL(i-1) \; \exp\left(-\frac{1}{ck}\right) + AL(i) \qquad 6.15$$

where PAL(i) is the animal pollution load in the channel at step i, SAL(i) is the accumulated amount in the sub-basin at time i, AL(i) is animal products generated in period i-1 to i, and ck is a decay coefficient for the particular pollutant. For the cases when there is runoff predicted, i.e. R(i)>0, the animal pollutant load is calculated as:

$$PAL(i) = SAL(i-1) \frac{R(i)}{MR} \qquad 6.16$$

and

$$SAL(i) = SAL(i-1) \exp\left(-\frac{1}{c_k}\right) + AL(i) - PAL(i) \qquad 6.17$$

where R(i) is the runoff in time i and MR is the runoff volume associated with the transport capacity. For convenience, it can be assumed that MR is 1.2 times the maximum runoff in a long data series.

The water quality variables considered in this model include Nitrogen, Phosphorus, and Biochemical Oxygen Demand (BOD). BOD is the amount of oxygen consumed by microorganisms while stabilising or degrading carbonaceous and nitrogenous compounds under aerobic conditions. The BOD value is widely used as an important indicator of the degree of water pollution. It is the primary regulatory tool used in limiting discharge of organic waste to water. Table 6.1 and 6.2 list the typical N, P, and BOD of excreta produced by livestock and from liquid wastes. The amounts of N, P and BOD produced per day by animals are summarised in Table 6.3.

Table 6.1: Typical BOD Levels (mg l^{-1}) produced by liquid wastes (Dodd, 1993)

Type of Waste	BOD (mg l^{-1})
Treated domestic sewage	20-60
Raw domestic sewage	300-400
Vegetable washings	500-3000
Dilute parlour and yard washings	1000-2000
Liquid wastes draining from slurry stores	1000-12000
Liquid sewage sludge	10000-20000
Cattle slurry	10000-20000
Pig slurry	20000-30000
Silage effluent	30000-80000
Milk	140000

Table 6.2: Amount and N & P contents of excreta produced by Livestock (Dodd, 1993)

Animal Type	Body weight (kg)	Typical slurry produced (l day^{-1})	Dry matter (%)	N (%)	P$_2$O$_5$ (%)
1 dairy cow	450-650	57	25	0.6	0.3
1 beef bullock	200-450	27	25	0.6	0.3
1 dry cow	90-120	4.0	25	0.6	0.3
1 sow + litter	90-120	12	25	0.6	0.6
1 pig, dry fed	45-75	4.0	25	0.6	0.6
1 pig, liquid fed	45-75	7.0	25	0.6	0.6
1 lamb	45	2.2			
1 mature sheep	60-80	4.0			
1000 laying hens	2000	115	70	2.4-4.2	2.2
1000 broilers+litter	1000	36	70	2.4-4.2	2.2
1000 turkeys+litter	5000	124	70	2.4-4.2	2.2
Cattle Slurry			10	0.5	0.2
Pig Slurry, dry fed			10	0.6	0.4
Pig Slurry, wet fed			6-10	0.5	0.2
Pig Slurry, Whey fed			2-4	0.3	0.2

Table 6.3: Amounts of N, P and BOD Produced by Various Animals

Animal Type	Designed weight (kg)	N (mg day^{-1})	P (mg day^{-1})	BOD (mg day^{-1} O$_2$)
1 Beef bullock	200-450	145800	72900	270000
1 Dairy cow	450-650	307800	153900	1140000
1 Sow	45-75	64800	64800	360000
1 Pig	45-75	21600	21600	80000
1 Sheep	45-75	21600	21600	48600
100 Chickens	200	248400	227700	270000
100 Ducks	200	248400	248400	270000
1 Horse	200-300	61236	30618	113000

Note: The weight density of animal dung is assumed to be 0.9 kg l^{-1}

6.4.6   River Water Quality Model

A river water quality model is a set of mathematical expressions defining the physical, biological and chemical processes that are assumed to take place in the water body. Given a set of river flow conditions provided by any of the runoff models described above, and wastewater discharges, the model will provide a set of outputs that include values for each of the water quality variables for each time period and location in the river system.

In the last two decades, as the water pollution problem has become more serious for many water management authorities, much attention has been paid to modelling and many water quality models have been developed (e.g. Loucks, 1976, 1981; Whitehead, 1984; De Smedt, 1989; DeVries and Hromadka, 1992; James *et al.,* 1993). The simplest approach is due to Streeter and Phelps who, in 1925, developed the first mathematical model of a river reach to describe the balance of dissolved oxygen, and to predict the biochemical oxygen demand (BOD) and dissolved oxygen (DO) concentrations or deficits resulting from the discharge of biodegradable organic waste into the river system. The Streeter-Phelps equation has the following form:

$$\frac{dD}{dt} = k_1 L - k_2 D \qquad\qquad 6.18$$

where L is the carbonaceous biochemical oxygen demand and D is the oxygen deficit.

This comparatively simple formulation states that the change in the dissolved oxygen deficit from the saturation value (i.e. 0) in a river with a steady discharge can be represented by the sum of two processes: (1) oxygen removal from the water by the chemical and biochemical oxidation of dissolved and suspended organic matter, and (2) oxygenation of the water from the air through the air-water surface. It provides an analytical solution describing the dissolved oxygen deficit as a function of time from the point of discharge of a steady source of pollutant. It provides a means of modelling the transport downstream in the channel network of the effects of the contamination carried by runoff from the catchment and estimated by the catchment models. More recently with the development of computers and computation techniques, methods have been developed for solving large sets of simultaneous algebraic equations and finite-difference representations of more complex linear and nonlinear differential equations. However, most of the water quality models in actual use today are extensions of the Streeter-Phelps model.

The basic principle of river water quality modelling is the conservation of mass. In the generalised case of a river system, a river reach of length X can be characterised by lateral inflow q (discharge per width) and upstream inflow Q_0 and tributary discharge Q_i. The initial concentration C, of each component dictates the quality of water in the reach and flowing into the next reach downstream. If it is assumed that the water quality is only influenced by conservative chemicals (i.e., the constituents are not influenced by any chemical, physico-chemical or biochemical processes once in the river water), then the total mass dissolved in the river flow remains constant. If it is also assumed that the river flows and concentrations are steady, then the resulting concentration C for a given water quality variable can be calculated using the following material balance equation:

$$QC = Q_0 C_0 + \sum_{i=1}^{n} Q_i C_i + XqC_l \qquad 6.19$$

and

$$C = \frac{Q_0 C_0 + \sum_{i=1}^{n} Q_i C_i + XqC_l}{Q_0 + \sum_{i=1}^{n} Q_i + Xq} \qquad 6.20$$

By applying the same mass balance to a river reach which has an additional point source mass input M (mass per time) of soluble waste then the resulting concentration C can be computed as:

$$C = \frac{Q_0 C_0 + \sum_{i=1}^{n} Q_i C_i + XqC_l + M}{Q_0 + \sum_{i=1}^{n} Q_i + Xq} \qquad 6.21$$

This approach allows the DSS to estimate the concentration of whatever chemical species is being studied at specified points along a channel. Thus the DSS has (i) models for estimating runoff (and its sediment and chemical content) from sub-basins into the channel network and (ii) models for estimating how these are carried further downstream in the channel network.

The simple advective model described in the last section is based on the assumptions of conservative substances and steady state conditions. It is called advective because the pollutants are assumed to be carried along at a speed equal to the average velocity of the water. However more complex models include many other processes, including: (i) dispersion which is non-advective transport due to the migration of a solute in response to a turbulent diffusion and non-uniformity of velocity and concentration gradient in the water and (ii) decay where the pollutant reacts with chemicals in the water, e.g.

oxidation, and change state or decay as they are carried along by the water. Models which account for the latter are called non-conservative models. Both of the above effects have been incorporated into the DSS.

There were two basic implementation options available when constructing the DSS. (1) The various models described above could be programmed in the macro (or built-in) language of the GIS package. This is AML for theARC/INFO package. Alternatively, (2) the models could be programmed in any standard third level programming language such as FORTRAN, PASCAL or BASIC and compiled into an executable module which could be called by the GIS but which did its calculations outside the GIS. The models required extensive calculations and these were generally accomplished faster by a stand alone external program than by the internal AML language. So the second option was chosen for the DSS. The GIS was programmed to prepare data files of the inputs required for each model then call the model program which stored its results in a file which was then read by the GIS package for storage, display and possibly further analysis.

6.5   Example application of surface water quality models applied to the River Dodder catchment, central-eastern Ireland

In order to illustrate how the models for deriving surface water quality predictions can be implemented, their application to the River Dodder catchment (Figure 6.5) in central-eastern Ireland are illustrated.

6.5.1   Using the hydrological model SMARD

The SMARD model has been tested using Dodder catchment data. The river system drains a catchment of 113 km^2 above Orwell Weir in Counties Dublin and Wicklow, Ireland. The main tributaries are the Little Dargle River I and the Owendoher River II to the east and the northern tributary V to the west (Figure 6.5). The upper part of the catchment consists of steep mountain slopes which are covered with a deep blanket of bog overlying granite rock and thin Quaternary (sub-soil) deposits. The lower part of the catchment area consists of flat undulating countryside intermingled with a considerable amount of urban development. Limestone bedrock is overlain largely by till deposits. The land-uses in the catchment are mainly urban and industrial units, urban parkland and sports greens, pastures and natural grasslands, agricultural areas, and peat bogs (Figure 6.6). Land-use data were derived from satellite remote sensing data. There are two drinking water supply reservoirs in the Glenasmole Valley part of the catchment.

For application of the distributed hydrological model (SMARD), the Dodder catchment was divided into five sub-basins according to its drainage system. As the confluences of the Little Dargle river and the Owendoher river to the River Dodder are near to the outlet of the catchment, only the discharge components flowing from the upper Dodder and from the northern tributary needed to be routed to the catchment outlet using the channel routing model.

There are 8 rainfall gauging stations in the Dodder catchment. Data were interpolated across the catchment area using Thiessen polygons and the total rainfall over the whole catchment for the period 1975 to 1990 was estimated as a weighted average (using the Thiessen polygon areas as weights) of the measured rainfall at each rain gauge. This approach was taken because there was insufficient data to apply a more complex interpolation, and data were incomplete so parameter adjustment was performed manually to achieve good model fits with available data.

6.5.2　Input data (boundaries, contours, land-use and soils) and parameter estimation in the GIS

The integration of the GIS with the distributed parameter hydrological model was dependent upon the availability of catchment boundary, soil association, topography, land-use, river system, and rainfall data. Soil association information was obtained from Gardner and Radford's (1980) map of the soils of Ireland. Elevation and river network information were assembled from information in Ordnance Survey maps and from site visits. Land cover was taken from the CORINE database, which is based on satellite imagery. Catchment boundaries were determined from the elevation data and observation.

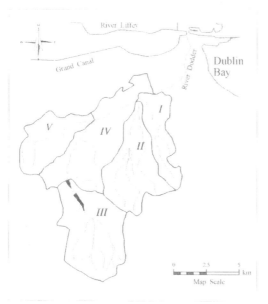

The SCS method, described above, gives tables which prescribe curve number (CN) values for each different soil types and land use. The mean value of CN in a sub-basin was calculated using the aerial weighted algorithm. The sub-catchment areas, river lengths and slopes were determined using the digitised data processed by GIS commands. The estimated parameter values are summarised in Table 6.4.

Figure 6.5: The River Dodder catchment (Little Dargle River I: 11.80 km², Owendoher River II: 22.49 km², Upper Dodder III: 27.99 km², Lower Dodder IV: 35.82 km², Northern Tributary V: 15.23 km².)

The simulated hydrograph results for the years 1993 and 1994 for the Dodder catchment match the measured data reasonably well, especially the peaks (Figure 6.7). This was achieved without any model calibration. However, the simulated low flows were generally smaller than those

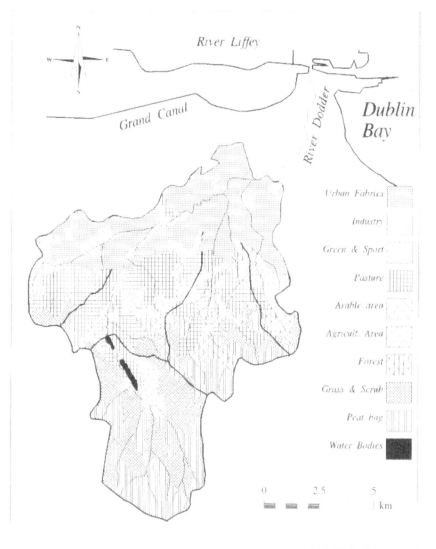

Figure 6.6: Land-use Classification in the Dodder Catchment (from CORINE).
Areas in km²: Urban fabric: 34.886; Industrial and commercial unit: 1.996;
Green park and sport area: 2.953; Pasture: 22.641; Non-irrigated arable area:
1.339; Agricultural area: 11.532; Forest: 7.414; Natural Grass and Scrub:
17.052; Peat bog: 15.761; Water body: 0.413.

Table 6.4: Estimated Parameter Values in the Dodder Catchment

Sub-Basin				Parameter			
Number	Name	Area	L	L_{ca}	S_{1085}	URBT	CN
I	Dargle	11.8	7633.2	3502.5	0.02764	1.527	64.4
II	Owendoher	22.5	8995.8	5268.7	0.03691	1.126	78.9
III	Upper Dodder	28.0	9220.9	4681.5	0.04447	1.000	82.2
IV	Lower Dodder	35.8	11284	5170.4	0.00620	1.505	76.8
V	Northern Stream	15.2	6815.2	3788.6	0.01981	1.462	77.4

L = channel length (m) L_{ca} = channel length from outlet to point nearest catchment centroid (m)
S_{1085} = channel slope between 10 and 85 % distance from the outlet CN = curve number
URBT = 1 + proportion of catchment urbanised A = area (km^2)

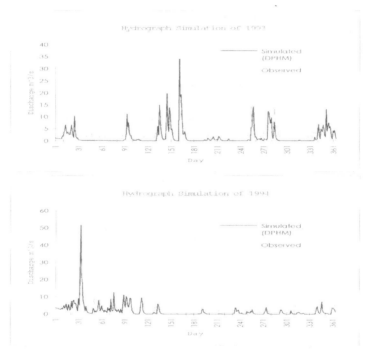

Figure 6.7: Simulated and measured hydrographs for the Dodder catchment, 1993 and 1994.

observed. This is because the version of SMAR used did not have a slow flow component suitable for modelling the gradual subsurface drainage of water which provides a low level of "base" flow even on dry days. The low flows are not critical to the purpose of the DSS since soil erosion and nutrient wash off mainly occur during extreme rainfall events and these are modelled well by SMAR. Nevertheless, since this work was done a groundwater component has been added to the SMAR model and will be incorporated into the DSS in future.

The data required for the erosion and nutrient loss estimates are assembled as follows. The soil erodability factor was estimated from the soil type data. There were 6 soil associations identified in the Dodder catchment. A soil erodibility factor K, based on US SCS data and the Universal Soil Loss Equation (USLE), was assigned to typify each soil association Table 6.5). The values of crop management factor C (also based on the USLE factors) were determined using the CORINE land use data (Table 6.6). The livestock distribution data were extracted from the *Census of Agriculture of June 1991* (Central Statistics Office of Ireland, 1994), for used in the model (Table 6.7).

Table 6.5: Values of Soil Erodibility Factor K in the Dodder Catchment

Soil Association No.	Grain Size Distribution	K tons/km^2
1	Sand 60%, Silt 38%, Clay 2%	59.31
4	Sand 70%, Silt 22%, Clay 8%	59.31
8	Sand 70%, Silt 21%, Clay 9%	59.31
9	Sand 33%, Silt 38%, Clay 29%	61.78
38	Sand 44%, Silt 32%, Clay 24%	61.78
40	Sand 26%, Silt 38%, Clay 36%	79.09

K=0 for urban areas

Table 6.6: Values of Crop Management Factor C in the Dodder Catchment

Land use	C	Note
Green land, Pastures	0.013	for 80% ground cover
Forests	0.012	for 80% ground cover
Grass and scrubs	0.011	for 80% ground cover
Agricultural areas	0.060	straw
Peat bogs	0.100	for 40% ground cover

Table 6.7: Livestock Number per km^2 in the Dodder Catchment

Sub-basin	I	II	III	IV	V
Beef Bullock	52.2	52.2	52.2	0	39.7
Dairy Cow	11.2	11.2	11.2	0	5.7
Sow	0.6	0.6	0.6	0	0.5
Pig	6.7	6.7	6.7	0	8.8
Sheep	255	255	255	0	86.5
100 poultry	0.83	0.83	0.83	0	0.38
Horse	1.0	1.0	1.0	0	1.8

6.5.3   Initial runs to evaluate the performance of the GIS based model

Initial runs of the model indicated that soil erosion should not be a serious pollution source in the Dodder catchment. These used daily rainfall data for the year 1994 and the estimated parameter values described above. Soil erosion was only predicted to occur on

2 days (Feb. 2 and 3) when daily rainfall was about 40-60 mm. There are two main reasons why the model estimated so little soil erosion in this catchment:

1. **Very good land cover.** From the statistics of the Land-use coverage the pastures, green lands, forests, grass and scrub cover about 63%, urban fabric and industrial units cover about 27%, and agricultural areas only cover about 10%. Therefore, the available soil for erosion in the catchment is quite low.

2. **Rainfall Data interval.** Many studies have shown that the rainfall intensity, especially the maximum 30-min amount of rainfall is the most important factor in determining the USLE model sensitivity (c.f. Shen, 1992). However, since only daily rainfall data was used in the initial runs the model could not take account of higher intensity and shorter duration showers which are likely to cause the most erosion. Soil erosion estimated from daily rainfall is likely to be smaller than the actual value.

The rainfall data were converted into a series of 1 hour intensity values using an algorithm which preserves the observed daily rainfall total and yet has the intensity duration relationship developed in the Flood Study Report Method (Natural Environment Research Council, 1975). For days in which there is rain, this method gives an hourly rainfall sequence which has some hours with intensities greater than the daily average and some below it. Using this modified data, the predicted soil erosion was much greater and closer to the values expected (Table 6.8). As yet there are insufficient measured data to compare with these calculations so this component of the model is not yet fully tested. The data required could only be obtained by a very intensive and long-term measuring campaign.

Table 6.8 shows quite a difference in estimated soil loss between the sub-basins with III having the greatest loss. This is the steepest sub-basin and also because of its location and altitude receives a greater depth of rainfall than the others. Both factors contribute to the greater erosion rates. In contrast, sub-basins I, IV and V are the flattest and lowest and so have the least rainfall and these show considerably less erosion.

Table 6.8: Annual Soil Erosion (tons) in Dodder Catchment (using 1-hour Rainfall Intensity data for 1994)

	Sub-Basin				
	I	II	III	IV	V
Jan.	23.9	4621.9	12894.4	979.6	154.2
Feb.	151.5	11453.8	64800.3	1758.6	765.8
Mar.	26.3	2139.4	12869.5	324.9	156.1
April	38.7	3079.1	20608.8	488.9	223.5
May	66.3	1934.7	5693.1	410.8	91.3
June	.0	17.0	1023.1	.0	7.3
July	17.5	662.9	2939.5	134.1	41.5
Aug.	10.7	1089.2	9576.5	150.6	102.6
Sept.	50.0	2216.9	8863.3	434.1	131.1
Oct.	10.8	1127.3	8775.7	161.1	95.2
Nov.	54.6	2166.5	8479.7	399.7	125.2
Dec.	57.9	5135.8	28142.4	656.5	364.1

6.5.4   Water Quality Component

The water quality modelling capability of the DSS is illustrated here with an hypothetical example for the lower Dodder river, i.e. the reach from the dam of the lower Glenasmole reservoir to the Orwell weir. The reach length is 11.3 km with an average slope of 0.62%. It receives the flows and pollutants from the Little Dargle river, the Owendoher river, the upper Dodder river, the northern tributary, and its own sub-basin. It also receives the industrial and urban sewage from the highly developed urban areas distributed on both its banks. The hydrographs and non-point source pollution from each sub-basin were estimated using the models described above.

In the DSS, the water quality variables considered are BOD, Nitrogen, and Phosphorus. They are the water quality indices of most concern in the lower Dodder river (Flanagan, et al, 1986, 1992). The mean river width was taken to be 15 m, and the water depth was calculated using the estimated discharge at the Orwell weir and the rating curve given in the Dodder River Flood Study (Hennigan et al., 1988).

The DSS has two type of water quality transport model, the simple conservative model, based on conservation of mass, described above and a more complex model which takes account of the decay of the pollutant as it is carried with the flow. Since insufficient measured data was available to calibrate the model parameters, these were estimated from a visual examination of the river.

For the purposes of testing it was hypothesised that there were 2 major effluent discharges on the right bank and another 2 effluent discharges on the left hand bank of the lower Dodder river. Table 6.9 gives the assumed wastewater data of these effluents. Using the data and model parameters discussed above as inputs to the river water quality model, concentration hydrograph (change with time) for each river section, and the concentration values at each section (change along the river) for any specified time were estimated. For example, Figure 6.8 illustrates the variations of flow and concentration along the river reach at a specified time. It is possible to identify the water quality situation in each small river reach, and find where are the main pollution sources, in order to take measures to reduce or control the pollution. Although there were insufficient data for a proper calibration this test of the model, estimated BOD values within the range of measured spot values, a maximum BOD 8.9 mg/l and a median of 2.7 mg/l at the outlet of the catchment for the period 1987 to 1990 (Flanagan and Larkin, 1992). However, much more data on inputs and river response with sufficient spatial and temporal resolution are highly desirable to further test the model.

Table 6.9: Hypothetical Effluent discharges to the Lower Dodder River (Assumed Data)

Section	Discharge $m^{-3} s^{-1}$	Content N $mg\ l^{-1}$	Content P $mg\ l^{-1}$	Content BOD $mg\ l^{-1}$
Effluent-1	0.1	50	20	10
Effluent-3	0.1	100	25	15
Effluent-4	0.3	30	10	20
Effluent-6	0.1	10	20	10

6.5.5   Summary

The inputs to any of the models can be changed by the user and the effect of the change estimated. For instance, the land-use category of any area can be changed and its effects on both the quantity and quality of runoff estimated. Similarly, a point load of effluent can be added at any point and its effects on the downstream water quality estimated. This could be used to estimate the effects of an accident and sudden release of pollutant at any given location or of a more permanent point discharge. Additional components and water quality parameters can be added to the model as required.

6.6   The Future for GIS Applications in Hydrology

The application of GIS in hydrology and water-related fields should continue and make use of all new developments in terms of both ideas, models and software. However, the future of GIS applications for hydrological modelling is not obviously evident. It is important to consider where things are going and where they should go. The following suggestions for future development are made with the intention of stimulating the reader with the exciting possibilities the future holds.

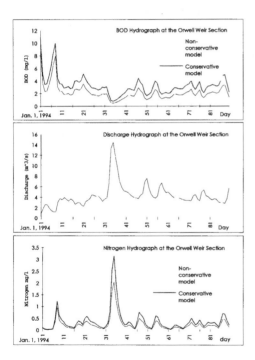

Data collecting methods. To enhance the application of GIS, more remote sensing data are required. Although remote sensing has been possible for many years, data derived from this source are not directly used in many hydrological models. Remote sensing has a significant potential to add to hydrological modelling through quantification of land use and other geographic information of the catchments being studied. Remote sensing is an effective method of collecting the type of data required for quick assessments and real-time decision support, especially in a large scale catchment or region. With the advent of new, high resolution platforms working on a commercial basis, the supply of data will be guaranteed and there is much scope for finding the best ways of exploiting these opportunities.

Figure 6.8: Flow and BOD Concentration (obtained using non-conservative model) of Jan. 8, 1994

Sophistication of hydrological models. Hydrological models play a key role in applications. The facilities of GIS for inputting and analysis of geo-referenced data will continue to encourage hydrologists to invent more sophisticated and effective hydrological models, and develop more effective methods or tools to estimate model parameters again using GIS. The current new-frontiers in hydrological model studies are: spatially distributed catchment properties, partial area flow, surface water-groundwater interaction, regional and global hydrology and spatial patterns of drought (Maidment, 1993). There is no doubt that the application of GIS will facilitate an increase in sophistication of hydrological models, and strengthen the physical basis of hydrological models.

The widespread application of GIS in hydrological modelling, will create a need for more sophisticated GIS tools. The further development of GIS should at least embrace the following aspects:

More object-oriented GIS systems. Up to a few years ago, most professional GISs only operated with their own special programming environment. This greatly limited the application of the developed hydrological software. Therefore, more object-oriented self-developed systems, which can be used without any special requirement and linked to any programme written in an object-oriented language such as C++, Java, Delphi or Visual Basic are necessary. This is the way many GIS systems are now developing.

More sophisticated spatial interpolators. With the development of automated monitoring technology there is an increasing amount of data available to the end-user. There will always be a need for interpolation of this data and as geostatistical techniques mature and develop, they can be implemented in GIS systems (e.g. ArcView now comes with semivariogram analysis and kriged interpolation as standard).

More specialist systems. From the view of practice, more specialist systems are required. It is not necessary that they include everything aspect of hydrology, but they could be developed to solve one or several practical hydrological problems. Current professional GISs, could develop some special tools or sub-systems for hydrological modelling so as to enhance their utility. There is much to be done!

References

Abbott, M.B., Bathurst, J.C., Cunge, J.A., O'Connell, P.E., and Rasmussen, J. (1986a): Introduction to the European hydrological system - System Hydrologique Europeen, `SHE',1: History and Philosophy of a physically-based, distributed modelling system. *Journal of Hydrology* **87**: 45-59.

Abbott, M.B., Bathurst, J.C., Cunge, J.A., O'Connell, P.E., and Rasmussen, J. (1986b): Introduction to the European hydrological system - System Hydrologique Europeen, `SHE',2: Structure of a physically-based, distributed modelling system. *Journal of Hydrology* **87**: 61-77.

Bender, M. and Simonovic, S. P. (1994): Decision-Support System for Long-Range Stream Flow Forecasting. *Journal of Computing in Civil Engineering* **8**: 20-33.

Bruen, M. and Masopha M. (1991): Use of a GIS to investigate the potential threat to groundwater of non-point source agricultural pollution in Ireland. *Proceedings of the First Irish Environmental Engineering Conference*, pp.189-196.

Camara, A. (1991): Decision support system for estuarine water -quality management. *Journal of Water Resources Planning and Management* **116**: 417-32.

Central Statistics Office. (1994): *Census of Agriculture, June 1991*. Government Publications Sale Office. Dublin.

Chow, V.T. (1964): *Handbook of Applied Hydrology: A Compendium of Water Resources Technology*. New York, McGraw-Hill.

Commission of the EC. (1989): *CORINE PROGRAMME: Land Cover Project*. Brussels.

Costa, F.S., Damazio, J.M., Das Neves, F.P. and Simabuguro, M. de F. R.. (1996): Linking a synthetic storm generation model with the IDRISI GIS. Application of GIS in Hydrology and Water Resources Management, *IAHS Publication* **235**: 107-13.

Costa, J.R., Jesus, H.B. and Lacerda, M. (1996): Intergrating GIS and time series analysis for water resources management in Portugal. Application of GIS in Hydrology and Water Resources Management, *IAHS Publication* **235**: 289-298.

Cunge, J.A. (1969): On the subject of a flood propagation computation method (Muskingum method). *Journal of Hydraulic Research* **2**: 205-30.

Davis, J.R., Nanninga, P.M., Biggins,J. and Laut, P. (1991): Decision support system for analyzing impact of catchment policies. *Journal of Water Resources Planning and Management* **117**: 399-414.

Department of Hydrology, MWR, China. (1992): *Water Resources Assessment for China*. China Water and Power Press.

De Roo, P.J., Wesseling, C.G., Jetten, V.G. and Ritsema, C.J. (1996): LISEM: a physically-based hydrological and soil erosion model incorporated in a GIS. In Application of GIS in Hydrology and Water Resources Management, *IAHS Publication* **235**: 395-404.

De Roo, P.J. (1993): Validation of the ANSWERS catchment model for runoff and soil erosion simulation in catchments in the Netherlands and the United Kingdom. In Application of GIS in Hydrology and Water Resources Management, *IAHS Publication* **211**: 465-74.

De Smedt, F. (1989): *Introduction to River Water Quality Modeling*. VUB-press, Brussels.

DeVantier, B.A. and Feldman, A.D. (1993): Review of GIS application in hydrologic modeling. *Journal of Water Resources Planning and Management* **119**: 246-61.

DeVries, J.J. and Hromadka, T.V. (1992): Computer Models for Surface Water. In Maidment (Editor): *Handbook of Hydrology*. New York, McGraw-Hill Inc. pp.21 - 27.

Dodd, V. (1993): Environmental studies of agricultural impacts. Notes for the Environmental Hydrology continuing education course held at University College Dublin. (unpublished).

Dooge, J.I. (1986): Theory of Flood Routing. In: Kraijenhoff, D. A, and Moll, J. R. (Editors): *River Flow Modelling and Forecasting*. Dordrecht, Holland, D. Reidel Publishing Company. pp.39-66.

Dublin Corporation, Dublin, Kildare, Wicklow, and Meath County Councils. (1994): *Water Quality Management Plan for the Liffey Catchment*. Prepared in accordance with Section 15 of the Local Government (Water Pollution) Acts, 1977-1990.

Dunn, S.M., Mackay, R., Adams, R. and Oglethorpe, D. R. (1996): The hydrological component of the MELUP decision support system: an appraisal. *Journal of Hydrology* **177** : 213-35.

Eastman, J.R. (1990): *IDRISI - A Grid Based Geographic Analysis System*. Clark University, Graduate School of Geography, USA.

Environmental System Research Institute, Inc. (1993): *Understanding GIS, The ARC/INFO Method*. Esses, Longman Scientific & Technical.

Engle, B., Navulur, K., Cooper, B. and Hahn, L. (1996): Estimating groundwater vulnerability to non-point source pollution from nitrates and pesticides on a regional scale. In Application of GIS in Hydrology and Water Resources Management , *IAHS Publication* **235**: 521-26.

Engel, B.A., Srinivasan, R., Arnold, J. and Rewerts, C. (1993): Non-point source (NPS) pollution modeling using models integrated with geographic information systems (GIS). *Water Science Technology* **28**: 685-90.

Farisser, P. and Givone, P. (1993): Mapping and Management of Flood Plains. Application of GIS in Hydrology and Water Resources Management, *IAHS Publication* **211**: 485-90.

Fedra, K. (1993): Models, GIS, and Expert System: Integrated Water Models. Application of GIS in Hydrology and Water Resources Management, *IAHS Publication* **211**: 297-308.

Fedra, K. and Jamieson, D.G. (1996a): An object-oriented approach to model integration: a river basin information system example. Application of GIS in Hydrology and Water Resources Management, *IAHS Publication* **235**: 669-676.

Fedra, K. and Jamieson, D.G. (1996b): The `WaterWare' decision-support system for river-basin planning. 2. Planning capability. *Journal of Hydrology* **177**:177-98.

Flanagan, P.J., and Larkin, P.M. (Editors). (1986): *Water Quality in Ireland - The Current Position*. Part Two: River Quality Data. Dublin, An Foras Forbartha.

Flanagan, P.J., and Larkin, P.M. (Editors). (1992): *Water Quality in Ireland, 1987-1990*. Part Two: River Quality Data. Dublin, An Foras Forbartha.

Fread, D.L. (1992): Flow Routing. In Maidment, D.R. (Editor): *Handbook of Hydrology*. New York, McGraw-Hill Inc.

Frysinger, S.P., Thomas, R.P. and Parsons, A.M. (1993): Hydrological modelling and GIS: the Sandra Environmental support system. Application of GIS in Hydrology and Water Resources Management. *IAHS Publication* **211**:45-50.

Furst, J. (1993): Application of GIS in Decision Support Systems for Groundwater management. Application of GIS in Hydrology and Water Resources Management. *IAHS Publication* **211**: 13-22.

Gardiner, M.J. and Radford, T. (1980): *Soil Associations of Ireland and Their Land Use Potential*. National Soil Survey of Ireland, Dublin, An Foras Taluntais.

Guariso, G. and Werthner, H. (1989): *Environmental Decision Support Systems*. London, Halsted Press.

Gupta, S.K. and Solomon, S.I. (1977): Distributed numerical model for estimating runoff and sediment discharge of ungauged rivers. 1. Information system. *Water Resources Research* **13**: 613-18.

Heidtke, T.M. and Auer, M.T. (1993): Application of a GIS-based non-point source nutrient loading model for assessment of land development scenarios and water quality in Owasco lake. New York, *Water Science Technology* **28**: 595-604.

Hennigan, P., McDaid, J. and Keyes, J. (1988): Dodder River - Flood Study. Unpublished research report, Institution of Engineers of Ireland.

James, A. and Elliott, D.J. (1993): Models of Water Quality in Rivers. In: James, A. (Editor): *An Introduction to Water Quality Modelling.* John Wiley & Sons. Chichester. pp.141-181.

Jamieson, D.G. and Fedra, K. (1995): Use of a decision-support system for the restoration of the Rio Lerma in Mexico. In: *Proceedings of the International Congress on Modelling and Simulation (MODSIM 95)* **3**: 218-23.

Jamieson, D.G. (1996a): Preface (for the special issue on Decision Support Systems), *Journal of Hydrology* **177**: 161-2.

Jamieson, D.G. and Fedra, K. (1996b): The `WaterWare' decision-support system for river-basin planning. 1. Conceptual design. *Journal of Hydrology* **177**: 163-75.

Jamieson, D.G. and Fedra, K. (1996c): The `WaterWare' decision-support system for river-basin planning. 3. Example application. *Journal of Hydrology* **177**: 199-211.

Kachroo, R.K. (1992): River Flow Forecasting Part 5: Application of a conceptual model. *Journal of Hydrology* **133**: 147-78.

Kaden, S.O. (1993): GIS in Water-related Environment Planning and Management: Problems and Solutions. in Application of GIS in Hydrology and Water Resources Management, *IAHS Publication* **211**: 385-398.

Kamps, T.W. and Oltshoorn, T.N. (1996): Using GIS for hierarchial refinement of MODFLOW models. in Application of GIS in Hydrology and Water Resources. *IAHS Publication* **235**: 535-42.

Keser, G. and Bogardi, J. (1993): National water resources management planning based on GIS. In Application of GIS in Hydrology and Water Resources Management. *IAHS Publication* **211**: 439-48.

Khan, H. (1989): A comparative study of two conceptual rainfall-runoff models . Ph.D Thesis, Department of Engineering Hydrology, U.C.G, National University of Ireland.

King, N. (1996): Achieving decision support with GIS: Learning from water management applications in Sourth Africa. In Application of GIS in Hydrology and Water Resources Management. *IAHS Publication* **235**: 677-84.

Knisel, W.G. (Editor). (1980): *CREAMS: A Field Scale Model for Chemicals, Runoff, and Erosion from Agricultural Management System.* U.S. Department of Agriculture, Conservation Research Report No.26, 640 pp.

Lam, D.C.L., and Swayne, D.A. (1993): An Expert System Approach of Integrating Hydrological database, models and GIS: Application of the RAISON System. Application of GIS in Hydrology and Water Resources Management. *IAHS Publication* **211**: 23-34.

Lange W.J. de & van der Meij, J. L. (1993): A National groundwater model combined with GIS for Water Management in the Netherlands. Application of GIS in Hydrology and Water Resources Management. *IAHS Publication* **211**: 333-43.

Liang, G.C. (Editor). (1995): Proceedings of the 6th International Advanced Workshop on River Flow Forecasting. University College Galway.

Liang, G.C. (1986): Linear Models for Multiple Inflow - Single Outflow Flow Routing in Real Time. Ph.D Thesis, Department of Engineering Hydrology, U.C.G., National University of Ireland.

Lieste R., Kovar, K., Verlouw, J.G.W. and Gan, J.B.S. (1993): Development of the GIS-based `RIVM National Groundwater Model for the Netherlands'. Application of

GIS in Hydrology and Water Resources Management. *IAHS Publication* **211**: 641-50.

Loucks, D.P. (1947): Surface water quality management. In Biswas (editor): *Systems Approach to Water Management.* McGraw-Hill Inc. New York.

Loucks, D. P. (1981): Water Quality Models for River Systems. In Biswas, A.K. (Editor): *Models for Water Quality Management.* McGraw-Hill Inc., New York. pp.1-33.

Loucks, D.P. and Fedra, K. (1985): Interactive water resources modelling and model use: and overview. *Water Resources Research* **21**: 95-102.

Maidment, D. R. (1992): *Handbook of Hydrology.* New York, McGraw-Hill.

Maidment D.R. (1993a): GIS and hydrologic modeling. In Goodchild, M. F., Parks, B. O. and Steyaert, L. T. (Editors): *Environmental Modelling with GIS.* Oxford, Oxford University Press. pp.148-67.

Maidment D.R. (1993b): Developing a Spatial Distribution Unit Hydrograph by using GIS. Application of GIS in Hydrology and Water Resources Management. *IAHS Publication* **211**: 181-92.

Matthews, G.J., and Grabs, W. (1994): Hydro-economic development and tools for decision aiding. *Journal of Hydraulic Research* **32**: 15-19.

Meijerink, M.J., Mannaerts, C.M., De Brouwer, H.A. and Valenzuela, C.R. (1993): Application of ILWIS to Decision Support in Watershed Management: case study of the Keomering river basin. Application of GIS in Hydrology and Water Resources Management. *IAHS Publication* **211**: 35-43.

Michl, C. (1996): Using GIS, MODFLOW and MODPATH for groundwater management of an alluvial aquifer of the River Sieg, Germany. In Application of GIS in Hydrology and Water Resources Management. *IAHS Publication* **235**: 551-58.

Moore I.D., Gallant, J.C. and Guerra, L. (1993): Modelling the spatial variability of hydrological processing using GIS. Application of GIS in Hydrology and Water Resources Management. *IAHS Publication* **211**: 161-170.

Nachtnebel H.P. (1993): Application of Geographical Information Systems to Support Groundwater Modelling. Application of GIS in Hydrology and Water Resources Management. *IAHS Publication* **211**: 653-64.

Napolitano, P. and Fabbri, A.G. (1996): Single-parameter sensitivity analysis for aquifer vulnerability assessment using DRASTIC and SINTACS. Application of GIS in Hydrology and Water Resources Management. *IAHS Publication* **235**: 559-66.

Nash, J.E. (1959): A note on the Muskingum method of flood routing. *Journal of Geophysics Research* **64**: 1053-56.

Nash, J.E. (1960): An unit hydrograph study with particular reference to British catchments. *Proceedings of the Institute of Civil Engineers* **17**: 249-82.

Natural Environment Research Council. (1975): *Flood Studies Report.* London, Whitefriers Press

Nearing, M.A., Foster, G.R., Lane, L.J. and Finkner, S.C. (1989): A Process-Based Soil Erosion Model for USDA - Water Erosion Prediction Project Technology. *Transactions of the ASAE* **35**: 1587-93.

O'Connell, P.E., Nash, J.E. and Farrell, J.P. (1970): River flow forecasting through conceptual models, part-2. The Brosna catchment at Ferbane. *Journal of Hydrology* **10**:317-29.

O'Connor, K.M. (1992): Introduction. Special Issue of *Journal of Hydrology* 133: (vii)

Olivier, J.J., and McPherson, D.R. (1993): Application of GIS to Water Management in a Developing region. Application of GIS in Hydrology and Water Resources Management. *IAHS Publication* **211**: 417-26.

Paudyal,G., Hauno,K. and Olesen,K.W. (1995): A Flood Management System for Bangladesh - The strategy of the generic model-GIS connection. Proceedings of the International Conference on Realities of Floods - a multi-disciplinary review of flood management issues St. Louis - US Commission on Irrigation and Drainage.

Pundt, H., Hitchcock, A., Bluhm, M. and Streit, U.A. (1996): A GIS-supported freshwater information system including a pen-computer component for filed data recording. Application of GIS in Hydrology and Water Resources Management. *IAHS Publication* **235**: 703-711.

Recknagel, F., Beuschold, E. and Petersohn, U. (1991): DELAQUA - a prototype expert system for operational control and management of lake quality. *Water Science Technology* **24**: 283-90.

Reitsma, R.F. (1996): Structure and support of water-resources management and decision making. *Journal of Hydrology* **177**: 253-68.

Romanowicz, R., Beven, K., Freer, J and Moore, R. (1993): TOPMODEL as an application module within WIS. Application of GIS in Hydrology and Water Resources Management *IAHS Publication* **211**: 211-25.

Shen H.W. and Julien, P.Y. (1992): Erosion and Sediment Transport. In Maidment, D. (Editor): *Handbook of Hydrology*. New York, McGraw-Hill, Inc.

Simonovic, S.P. (1993): Flood control management by integrating GIS with expert systems: Winnipeg City case study. Application of GIS in Hydrology and Water Resources Management. *IAHS Publication* **211**: 61-73.

Singh, V.P. (1988): *Hydrologic Systems - Volume I: Rainfall-Runoff Modelling*. New Jersey, Prentice Hall.

Skelly, W.C., Henderson-Sellers, A. and Pitman, A.J. (1993): Land surface data: Global climate modelling requirements. In Goodchild, M.F., Parks, B.O. and Steyaert, L.T. (Editors): *Environmental Modelling with GIS*. Oxford University Press, pp. 135-141.

Somlyody, L. and Varis, O. (1992): Water Quality Modelling of Rivers and Lakes . International Institute for Appliied System Analysis (IIASA) unpublished working papers, WP-92-041.

Van Der Zel, D.W. and Rabe, F.V. (1993): APS on GIS: an operational forest hydrological GIS. Application of GIS in Hydrology and Water Resources Management *IAHS Publication* **211**: 541-545.

Vieira, J.R. (1994): Management Support System for aquatic environment - concepts and technologies. *Journal of Hydraulic Research* **32**: 161-81.

Walsh, M.R. (1993): Toward Spatial Decision Support Systems in water resources. *Journal of Water Resources Planning and Management* **119**: 158-169.

Whitehead, P.G. (1984): The Application of Mathematical Models of Water Quality and Pollution Transport: An International Survey. Project IHP-II-A.1.7.1, UNESCO, Paris.

Wishmeier, W.H., and Smith, D.D. (1978): *Predicting Rainfall Erosion Losses - A Guide to Conservation Planning*. USDA Agriculture Handbook 537.

Young, R.A., Otterby, M.A. and Roos, A. (1982): A technique for evaluating feedlot pollution potential. *Journal of Soil and Water Conservation* **37**: 21-23.

Young, R.A., Onstad, C.A., Bosch, D.D. and Anderson, W.P. (1989): AGNPS: A non-point-source Pollution Model for Evaluating Agricultural catchments. *Journal of Soil and Water Conservation* 44: 168-73.

Zhang, J.Y, O'Connor, K.M. and Liang, G.C. (1994): A software package for river flow forecasting based on the SMAR model. *Hydraulic Engineering Software* 1: 163-72.

Zhang, J.Y., Dowley, A. and Bruen, M. (1995): GIS and Its Application in Hydrology and Water Resources. *Advances in Water Science* 6: 190-6.

Zhang, J.Y., Dowley, A. and Bruen, M. (1996): Watershed runoff modelling and geographical information systems. Proceedings of the HydroInformatics'96 conference, September, Zurich.

Zhang, J.Y. (1996): A Decision Support System for Water Management at Catchment Scale based on a Geographical Information System. Ph.D. Thesis, National University of Ireland.

Chapter 7 Crop and Animal Disease Forecasting and Control–Regional Perspectives

P.C. Mercer[1], A. Bell[1], L.R. Cooke[1], L. Dowley[2], B. Dunne[2], T. Keane[3], T. Kennedy[2] and R. Leonard[2]

1. *Applied Plant Science Division, Department of Agriculture and Rural Development for Northern Ireland, Northern Ireland*
2. *Crop Protection and Breeding Department, Teagasc Oak Park Research Station, Carlow, Ireland*
3. *Met Éireann, Glasnevin, Dublin, Ireland (Retired)*

7.1   Introduction

From earliest times, it has been evident to farmers that there is a connection between animal health and crop growth on one hand and weather on the other. Droughts led to crop failure and starvation of herds, and conversely prolonged rain often led to rotting of crops and poor thriving of animals. Although the causal relationship was fairly obvious in the first situation, it was not so evident in the latter. Whereas rain may have been the predisposing factor, micro-organisms and internal parasites were often ultimately responsible for the demise of crops and death or debilitation of animals.

Even in the middle of the nineteenth century, at the time of the Irish Potato Famine, there was no general realisation of the role of micro-organisms in causing disease of plants or animals. Lindley (1845) was typical of the majority of scientists in that he considered that the initial outbreak of the potato disease in 1845 was directly caused by the "cold and cheerless summer" (Bourke, 1991). Although there were initially a few voices advocating a fungal cause (Berkeley, 1846; Morren, 1845), the theory of the weather as a direct cause gained ground and indeed the final report of an inquiry into the potato famine rejected "the theory of parasitic Fungi being the cause of the evil, which they refer(red) to atmospheric influences" (O'Neil, 1946). This theory held sway until an accurate life-cycle of the blight fungus, *Phytophthora infestans* was described by de Bary (1861; 1863). From that point onwards, the links between plant diseases and micro-organisms became firmly established.

However, neither this knowledge nor any link between plant disease epidemics and weather was of much practical significance until reliable methods of disease control were achieved with the arrival of the first fungicide, Bordeaux mixture (Millardet, 1885). In the first quarter of the twentieth century, fungicidal sprays became more widely applied and relationships between weather variables and plant disease spread began to be discovered, *eg*. Melhus (1915) in the USA showed a correlation between temperature and zoospore liberation of *Phytophthora infestans*. Simple models and disease-forecasting schemes followed soon afterwards. One such, for vine mildew (*Plasmopara viticola*), was in use in 1923 (Chaptal, 1923). A forecasting scheme for potato blight, based on cloudiness, presence of dew, rainfall and temperature was in use in the Netherlands in 1926 (van Everdingen, 1926). Others followed, in the UK

(Beaumont and Staniland, 1933) and in the USA (Crosier and Reddick, 1935). These schemes, when sufficiently robust, allowed a delay in the first application of fungicide. Although one of the main driving forces for such schemes today is a reduction in pesticide usage, both for environmental and cost reasons, the early schemes were more concerned with the phytotoxic damage caused by the copper fungicides, which could reduce yield if blight was not present (Large, 1959). Delaying the first application resulted in much less plant scorch.

Although the cause and effect of internal parasites, such as liver fluke, and animal health was established much earlier than for micro-organisms, an understanding of the effect of weather on such relationships is a relatively recent phenomenon. Olsen (1947) in the USA showed that part of the life cycle of the liver fluke is disrupted by heat and drought during the four summer months. Primitive forecasting systems based solely on general climatic conditions were later developed in Europe, e.g. Ollerenshaw and Rowlands (1959) and used to estimate the risk of infection.

Many of these relatively simple forecasting schemes for both plant and animal diseases are still in use with some modification. In recent years the advent of the computer has allowed the rapid analysis of much larger amounts of data and the ability to incorporate many more variables, such as crop cultivar, previous pesticide treatments, fertiliser regimes and seed rate (Hardwick, 1998). These have been incorporated into so-called Decision Support Systems (DSS) which should allow a farmer to key in crop and weather details and then receive advice on what pesticides to apply, when to apply them and at what rate (e.g. Paveley, 1998). Existing disease-forecasting schemes are primarily based on past weather, but eventually more sophisticated programs may become available which take into account predicted weather patterns.

The aim of this chapter is to outline experience in forecasting and modelling of crop and animal disease problems from an Irish perspective. Naturally such experience is the result of local problems – there is for example a large section on potato blight. However, many of the problems are also ones experienced in other parts of the world and findings of researchers from these areas are also drawn upon. The chapter is divided into two main sections, the first on crop problems, primarily those of potatoes and cereals, and the second, on problems associated with animals.

7.2   Crop Diseases

7.2.1   Potato blight forecasting in Ireland

Introduction
Late blight, caused by the fungus *Phytophthora infestans*, is the most devastating disease of the potato crop in Ireland and world-wide. The key to its success lies in the extreme rapidity of its spread. Sporangia are produced in vast numbers on infected potato foliage in warm, humid weather and are then dispersed by wind and rain. When they alight on a potato plant, if conditions are suitable, they infect the foliage and the pathogen quickly spreads through the tissue causing the leaves and stems to blacken and die. Spores may also be washed into the soil and infect the tubers. Blighted tubers

develop a characteristic dry rusty brown rot and are predisposed to secondary bacterial soft rot, which can lead to total crop loss. During the 1940s, it was recognised that, whereas there were always enough blighted plants to provide a primary infection source, blight was seldom widespread until later in the season. This led to attempts to determine the conditions needed for its spread and thus to define potential infection periods. *Phytophthora infestans* is only able to sporulate to cause infection when the weather is both sufficiently warm and humid or wet.

The first forecasting attempts in the UK were based on rules developed in the Netherlands (van Everdingen, 1926), but these failed to forecast blight sufficiently accurately. Subsequently, Beaumont formulated the Beaumont Period (Beaumont, 1947) which was later superseded by Smith Periods (Smith, 1956). An attempt to refine the system by Sparks (1980) was not successfully adopted, and the Smith Period system continues in use in the UK to the present day. However, alternative systems are currently being evaluated by the Ministry of Agriculture, Fisheries and Food (MAFF) in England and Wales and in Northern Ireland by the Department of Agriculture and Rural Development for Northern Ireland (DARD).

In Ireland, Bourke (1953) developed a set of rules for forecasting late blight which were first used in 1952, and are known as the 'Irish Rules'. These rules were based on experimental laboratory work carried out by Crosier in the USA (Bourke, 1955). The rules were used for the development of a late blight warning service that is run by Met Éireann (the Irish meteorological service).

Recent developments in information technology and the memory capacity of computers have made it possible to log weather data continuously for individual crops and to use this information in computer-based DSSs to predict the date of disease outbreak and to determine the most suitable intervals between sprays. The objective is to improve the economics of control - to optimise on dose and timing to achieve a balance between input and output. A short description of different systems used in Europe is given in Table 7.1.

Recent research on forecasting of potato plight has been given different emphases in Ireland and Northern Ireland. In Ireland, research has concentrated on evaluation (based on field trials) of different DSSs, in particular NegFry, while in Northern Ireland, it has been centred on empirical observations of the practical use of Smith periods.

Potato blight forecasting in Ireland
Over a three-year period, trials were conducted at Oak Park Research Centre, Carlow, Ireland. The trials compared routine spraying with two DSSs. These systems were NegFry and the Met Éireann blight warning service. NegFry is based on two methods from which it derives its name. The first is the Negative prognosis as developed by Ullrich and Schrödter (1966) which calculates the epidemic free period for *Phytophthora infestans* and then recommends the first spray at the end of this period. This is done by adding a series of 'daily risk values'.

Table 7.1: General overview of forecasting and Decision Support Systems

DSS	Country and e-mail	Original devel. year	Main users	Input	Output
ProPhy	Netherlands Info @Opticrop.nl	1988	Farmers Advisors Extension officers	Weather data Field data Other data	Weather overviews Field data Advice
NegFry	Denmark JensG.Hansen @agrisci.dk	1992	Farmers Advisors	Weather data Field data	1st spray timing Fungicide applications
Simphyt	Germany Bkleinhenz.lpp-mainz @agrarinfo.rpl.de	1982	Plant protection service, Extension officers	Weather data Field data	Field data Advice
Plant-Plus	Netherlands Plantplus @dacom.nl	1990	Farmers Advisors Suppliers Processors	Micro-climate Crop + product information	Disease maps Fungicide protection periods
Televis	Norway Arne.Hermansen @planteforsh.no	1957	Farmers Advisors	Weather data	Epidemio-logical data
Guntz-Divoux	France fredec.nord.pas-de-Calais @wanadoo.fr	1963	Advisors Extension service	Weather data	Advice line
Guntz-Divoux	Belgium Pcg @ping.be	1996	Advisors Extension service	Weather data	Advice line
I.P.I	Spain	1990	Advisors Farmers	Weather data	1st spray timing Epidemio-logical data
PhytoPR E+ 2000	Switzerland Hansrudolf.forrer @fal.admin.ch	1995	Advisors Farmers, Plant protection service	Weather data Field data Other data	Regional data Field data

(after Bouma and Hansen, 1999)

Accumulation of a risk value, for any given day, occurs only when the relative humidity is >90% and the temperature is between 8 °C and 24 °C. The actual value calculated depends upon the precise temperature when the relative humidity is ≥ 90%, with the greatest accumulations occurring when the temperature is in the ranges 10°C to 11.9 °C and 18.0 °C to 23.9 °C. The second half of the model is derived from a method developed by Fry *et al.* (1983) which calculates subsequent spraying intervals based on blight units. The blight units are calculated based on the length of any humid spell, the air temperature during the humid spell and upon the cultivar susceptibility to late blight. The NegFry model requires air temperature, relative humidity, rainfall, cultivar susceptibility and crop emergence date as inputs. The Met Éireann blight warning service follows the Irish Rules using data supplied by automatic weather stations to calculate the severity of blight spells and making use of synoptic weather charts to predict spells of blight weather and so give a spray warning. The spray warnings are issued over national television and radio as required during the growing season. The rules for forecasting late blight are as follows (Keane, 1986):

(a) A 12 h period with air temperature 10 °C or greater and relative humidity not less than 90%;
(b) Suitable conditions for free moisture to remain on the leaves for a further two hours or more;
(c) Effective blight hours (EBHs) begin on the 12th successive hour as in (a) if, to satisfy condition (b), there is precipitation between the 7th and 15th hour. Otherwise accumulated EBHs do not begin until the 16th hour;
(d) If two spells with blight conditions, the first as in (c) and the second as in (a), follow each other, within 5 hours or less between the ending of the first and the beginning of the next, no lead in period of 11 or 15 h need be deducted for the second spell.

The trials were performed in accordance with the European and Mediterranean Plant Protection Organisation (EPPO) guidelines for the biological evaluation of fungicides (EPPO, 1978) at Oak Park, using the cultivar Rooster, which has resistance ratings of 4 for foliage blight and 6 for tuber blight (1 - 9 scale; 1 – completely susceptible; 9 – totally resistant). Local weather data were provided by a Hardi Metpole (Hardi International) on-farm weather station. Results for the years 1996 to 1998 compared a 10-day routine treatment of mancozeb ('Dithane DF'), with mancozeb following the Met Éireann blight warnings, mancozeb following NegFry warnings and fluazinam ('Shirlan') following NegFry warnings.

The criteria used to compare the systems were delay in disease onset, area under the disease progress curve (AUDPC - a measure of the overall progress and severity of an epidemic), marketable yield and mass of blighted tubers. Tables 7.2 to 7.5 show the results for the three years for each criterion.

Met Éireann warnings. In both 1997 and 1998, the delay in disease onset due to the Met Éireann blight warnings was significantly less than the delay following 10-day routine spraying (Table 7.2). In 1996, the Met Éireann blight warnings gave a slightly, but not significantly, greater delay than the 10 day routine spray. The Met Éireann blight warning programme resulted in significantly higher AUDPCs than with the 10 day routine treatments in both 1997 and 1998 (Table 7.3). In 1996, the Met Éireann blight warnings spraying routine gave a slightly, but not significantly, lower AUDPC than the 10 day routine spray.

Table 7.2: Delay in disease onset (days) following the different fungicide routines from 1996 - 98

Treatment	1996	1997	1998
Unsprayed	0.0	0.0	0.0
Mancozeb – 10 day routine	12.8	24.5	59.8
Mancozeb – Met Éireann warnings	15.5	3.5	26.3
Mancozeb – NegFry	17.0	15.8	44.8
Fluazinam – NegFry	25.5	12.3	59.8
L.S.D (0.05%) (all treatments)	9.2	13.0	17.8
L.S.D (0.05%) (sprayed treatments)	9.8	15.4	13.9

Table 7.3: Area under the disease progress curve (AUDPC) following the different fungicide routines from 1996 - 98

Treatment	1996	1997	1998
Unsprayed	1783	2981	1522
Mancozeb – 10 day routine	215	364	0
Mancozeb - Met Éireann warnings	198	1506	19
Mancozeb – NegFry	33	778	1
Fluazinam – NegFry	11	0	0
L.S.D (0.05%) (all treatments)	403	504	1170
L.S.D (0.05%) (sprayed treatments)	244	499	18

In 1996 and 1998, the Met Éireann blight warning programme resulted in slightly, but not significantly better marketable yield and tuber blight control than the 10-day routine treatment (Tables 7.4 and 7.5). Conversely, in 1997, the 10 day routine treatment was significantly better than the Met Éireann blight warnings in terms of marketable yield and slightly but not significantly better in terms of tuber blight.

NegFry. In terms of delay in disease onset (Table 7.2), the only significant differences between the NegFry programmes and the 10-day routine were in 1996 and 1998. The NegFry programme with fluazinam in 1996 was significantly better than the 10-day routine programme, and in 1998, the 10-day routine programme and the NegFry programme with fluazinam were both significantly better than the NegFry programme with mancozeb. In respect of the AUDPC (Table 7.3), there were no significant differences between the NegFry and the 10 day routine treatment. There was only one significant difference in marketable yield (Table 7.4). In 1997, the NegFry routine with mancozeb produced a significantly lower marketable yield than the 10-day routine. The only significant difference in tuber blight was in 1996 between the two NegFry treatments. The mancozeb treatment was significantly worse than the fluazinam treatment.

Table 7.4: Yield of marketable tubers (t ha⁻¹) following the different fungicide routines from 1996 – 98

Treatment	1996	1997	1998
Unsprayed	39.2	29.7	24.5
Mancozeb – 10 day routine	47.8	47.9	34.2
Mancozeb - Met Éireann warnings	51.9	39.5	38.4
Mancozeb – NegFry	49.0	41.9	27.1
Fluazinam – NegFry	49.6	44.8	38.0
L.S.D (0.05%) (all treatments)	5.9	6.3	13.2
L.S.D (0.05%) (sprayed treatments)	6.2	5.6	13.3

Table 7.5: Yield (t ha⁻¹) of tubers infected with blight following the different fungicide routines from 1996 – 98

Treatment	1996	1997	1998
Unsprayed	0.01	0.19	0.10
Mancozeb – 10 day routine	0.05	0.29	0.04
Mancozeb – Met Éireann warnings	0.02	0.34	0.01
Mancozeb – NegFry	0.18	0.36	0.02
Fluazinam – NegFry	0.00	0.22	0.02
L.S.D (0.05%) (all treatments)	0.13	0.29	0.08
L.S.D (0.05%) (sprayed treatments)	0.15	0.34	0.05

Table 7.6: Number of sprays applied following the different fungicide routines from 1996 – 98

Spray routine	1996	1997	1998
10 day	9	9	10
NegFry	4	7	8
Met Éireann	3	5	4

Comparison of number of sprays applied. Table 7.6 shows the different number of sprays applied in each of the three years following the different programmes. The Met Éireann blight warning service always gave the lowest number of sprays, with the NegFry system having a number between the other two systems. The number of sprays applied in any one year depends upon the prevailing weather conditions. The year 1996 was average for rainfall, humidity and temperature, and so the level of spray application in that year could be considered to be about average. The years 1997 and 1998 were both wetter than average during the growing season with high relative humidity and cool temperatures. It is therefore possible that, on average, NegFry will indicate only around 5 sprays per year. This is around a 50% saving over a 10-day routine spray programme.

The results at Oak Park show that in an average blight year (1996), the NegFry system can reduce spray frequency (standard rates) by 40% without any loss in disease control. The best results were achieved with full rate fluazinam indicating that effectiveness of the system may be fungicide dependent. In a very severe blight year such as 1997, the reduction in fungicide use was 25%. This confirms that a DSS can have a positive effect in reducing fungicide inputs for potatoes under Irish conditions without loss in disease control. However, reduced rates in combination with reduced frequencies of application were found to be ineffective.

The blight warnings, as issued by Met Éireann, are very effective at identifying periods of high blight pressure and can trigger extra sprays. However, sole dependence on warnings issued by Met Éireann was found to give less effective blight control than routine spraying. The years where the Met Éireann warnings failed coincided with years of high blight pressure, and a failure to control blight in these years could result in the loss of the entire yield of tubers. These results are similar to earlier results (Frost, 1994) which found that the Met Éireann warnings failed to give adequate control in some 30% of trials. In terms of practical use, this means that it is possible to follow a NegFry spraying routine exactly and obtain good blight control. The Met Éireann blight warning service predictions can be used as a guide to adjust a routine programme, and to suggest when to start a routine spray programme.

On-going work. Work is continuing at Oak Park to compare the operation of different DSSs, as part of an EU-wide project, with the aim of seeing how the different models compare under different weather regimes. The systems under evaluation are NegFry, Plant Plus, ProPhy and Simphyt.

Potato blight forecasting in Northern Ireland
In Northern Ireland in most years, weather conditions during the June-September period favour moderately severe or severe outbreaks of potato blight. DARD is responsible for issuing Potato Blight Warnings to growers. The objectives of the current forecasting system are: (1) to indicate to growers when to start spray programmes and (2) to inform growers when to modify spray interval. These rather limited objectives are dictated by several considerations. First, the current system covers the whole of Northern Ireland, but relies on meteorological data from a single site, so that the accuracy of the system cannot justify more detailed advice to growers. Second, the fungicides approved for potato blight control in the UK have limited curative activity. Curative use is generally not recommended by agrochemical companies due to the potential for crop loss if applications are made too late, and also because of concerns that curative use may encourage selection of fungicide resistant pathogen strains. This severely restricts the grower's window of opportunity to respond to a forecast risk of infection. Finally, the frequent and extended periods of unsettled and wet weather which occur in the growing period in Northern Ireland make it difficult for growers to apply sprays following a risk prediction.

The Smith Period system has been used for over twenty years in Northern Ireland, as elsewhere in the UK. A Smith Period is defined as two consecutive 24 hour periods, starting at 9 a.m. GMT, 10 a.m. British Summer Time (BST), in which the minimum temperature does not fall below 10 °C and there are at least 11 hours of 90% or greater relative humidity (RH) on each day (Smith, 1956).

Before 1993, temperature and humidity data were provided by the Meteorological Service from the main Northern Ireland Meteorological Station at Belfast International Airport, Aldergrove. However, the increasing cost of data dictated a switch to the use of an automatic meteorological station based at Newforge Lane, Belfast. The two systems were operated in parallel for one season with good agreement and thereafter only data from the automatic station were used. The incidence of Smith Periods in Northern Ireland in the last five years has recently been compared with the occurrence of blight outbreaks as reported by the DARD Potato Inspectors and Advisers.

In recent years, as in Ireland, DSSs, such as Plant Plus, have been evaluated (by the DARD Agri-food Development Service) using on-farm weather stations.

Use of Smith Periods: inputs of meteorological data. Meteorological data (temperature and humidity) are provided by an ELE automatic meteorological station sited at Newforge Lane, Belfast adjacent to an official meteorological station (Stevenson Screen). Readings are taken every 10 min. and converted to hourly records, which are off-loaded daily between June and September shortly after 10 am BST. The data are processed on a PC using ELE Dialog software and used to calculate Smith Periods. Some empiricism is used when deciding whether to issue a Blight Warning if conditions fall just outside strict Smith criteria in terms of either minimum temperature or hours of 90% RH The readings are compared with those from the official meteorological station as a quality control.

Inputs: details of blight outbreaks from DARD inspectors and advisers. DARD Potato Inspectors and Advisers report all early outbreaks of blight in Northern Ireland to Applied Plant Science Division (APSD) and these are used in Potato Blight Warnings. Potato Inspectors are also requested to provide samples of potato blight from crops in their areas (for use in the annual survey of phenylamide resistance in *Phytophthora infestans*) and the distribution of these provides an indication of the occurrence of infection.

Outputs: blight warnings. Potato Blight Warnings (occurrence of Infection Periods) are issued to growers in Northern Ireland via '*Blightline*', a 24 h recorded telephone information service on potato blight operated by APSD staff between June and August each year. *Blightline* is up-dated approximately twice weekly and, as well as providing details of Infection Periods, includes information on recent blight outbreaks and topical advice on blight control. Blight Warnings are also given via local radio (e.g. Radio Ulster's daily *Farmgate* programme) and more general reports of blight risk through press releases issued to the local farming papers.

Occurrence of Smith Periods in Northern Ireland, 1993-1998. Smith Periods usually occur from late May onwards in Northern Ireland (Table 7.7). Before this, night temperatures usually fall well below 10°C and thus the minimum temperature requirement prevents the earlier occurrence of Smith Periods. However, in 1997 there was a near Smith Period over the three days 17-20 May when the number of hours with over 90% RH totalled 69, but the minimum temperature fell to 9°C and, in 1998, Smith Periods were recorded 12 - 14 May and also 22 - 25 May.

Table 7.7: Occurrence of Smith Periods and initial blight outbreaks, 1993-98

Year	First Smith Period (A)	First blight report (B)	Time (days) A to B
1993	29-31 May	1 June	1
1994	19-21 June	30 June	9
1995	27-31 May	20 June	20
1996	9-11 June	28 June	17
1997	17-20 May	30 May	10
1998	12-14 May	8 June	25

In most years, sporadic Smith Periods were recorded during June, but the bulk of Infection Periods occurred during July and August and continued into September. Statutory haulm destruction dates are set for seed potato crops in Northern Ireland usually before the end of August. After this, only foliage of ware crops may be infected by blight and by late September most of these are either burnt down or senescent.

Relationship between Smith Periods and initial blight outbreaks, 1993-1998. Initial blight outbreaks are most frequently observed in field crops in Northern Ireland, although the occasional infected potato dump is reported. In each of the years 1993-1998, Smith Periods occurred before blight was seen. In 1994-1998, the time which elapsed between the first Smith Period and the first reported blight outbreak varied from 9 to 25 days (Table 7.7). In 1993, blight was seen in one field (and on a dump in a different location) only one day after the first Smith Period, so this infection must have pre-dated the Smith Period.

It is harder to relate Smith Periods to blight incidence within seasons as indicated by blight reports and samples received. This may be partly because sampling later in the season is not necessarily a reliable indicator of blight incidence. The Potato Inspectors also provide estimates of the number of crops with >5% blight at haulm destruction, but this does not appear to be related to Smith Periods (Table 7.8).

Although one of the objectives of the current Northern Ireland blight forecasting system is to advise growers when to modify the spray interval, the warnings issued indicate only the risk of infection (low, moderate, high, extremely high). No data are available regarding the extent to which growers use this information in deciding how often to apply fungicide sprays. Applying fungicide sprays in direct response to Blight Warnings (other than at the start of the season) is problematic due to the difficulty of accurately timing sprays in Ireland's unsettled summer weather. Given these limited objectives, Smith Periods have performed reasonably well as a predictor of the first field outbreaks of blight and providing a start date for spray programmes. Within seasons, comparison of the occurrence of Smith Periods with reports of blight outbreaks suggests that Smith Periods tend to over-estimate blight risk, particularly in a season such as 1995 when there was high humidity, but relatively little rain. This is perhaps not surprising, as Smith Periods are not intended as a forecasting system for use within seasons, although they are often used that way.

Table 7.8: Number of Smith Periods and foliage blight incidence, 1993-98

Year	No. of "Smith days"	No. of Infection Periods	No. of days in Infection Periods	Seed crops (%) with >5% blight
1993	19	5	12	13
1994	35	7	25	8
1995	46	10	37	0
1996	58	10	53	3
1997	69	13	66	11
1998	37	9	26	8

On-going work. Field-scale demonstrations of the DSS Plant-Plus, developed in the Netherlands (Hadders, 1996) are currently being conducted by DARD's Agri-food Development Service in conjunction with commercial companies. From 1998-2000, crops were subdivided into two areas which were sprayed with either a conventional programme, with the start date dictated by DARD's Blight Warnings, or with a programme derived from interpretation of the output of Plant-Plus. In each year, both programmes completely controlled infection and no symptoms developed in either area. Spray numbers were similar with the two programmes, probably because conditions were extremely favourable to infection.

Conclusions and future work
Further developments of these systems will integrate the use of weather forecasts, to predict the occurrence of weather favourable to blight. Model builders are also working on Internet-based programs that will allow farmers to log-on to get individual spray timings. However, while farmers in Ireland would like to reduce the number of sprays applied to the potato crop in any given year, they also demand effective control of blight. Blight can not only reduce the yield but may also reduce tuber quality and sale value. DSSs for late blight control must achieve either as good control as routine spraying, but with fewer applications, or improve control by optimising application timing. Such systems must be fully evaluated to ensure that they work under Irish conditions, neither predicting too early and wasting spray, nor predicting too late and so not controlling blight effectively.

7.2.2   Forecasting of cereal diseases

Research in Northern Ireland
The main cereal crops in Northern Ireland are spring barley and winter wheat and the main diseases which affect them are *Erysiphe graminis* (mildew) and *Septoria tritici* respectively. Various attempts have been made in Britain and elsewhere to forecast these diseases (Polley and Smith, 1973, Parker *et al*, 1994) as they are both weather-dependent, mildew tending to proliferate in warm, dry conditions and *Septoria tritici* in wet conditions.

As there is generally moderate rainfall during the main part of the growing season in Northern Ireland (Figure 7.1), *Septoria tritici* is usually present at epidemic proportions, although its severity is also partially dependent on cultivar resistance. Mildew is most prolific under warm, dry conditions and for that reason tends to be worse in the drier south-east than in the wetter north-west (Figure 7.2). Generally, mildew has been the

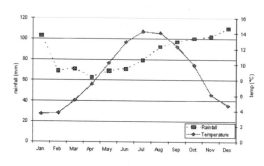

Figure 7.1: Mean monthly rainfall (mm) and temperature (^{0}C) in Northern Ireland from 1951-1980 (Anonymous, 1951-1980)

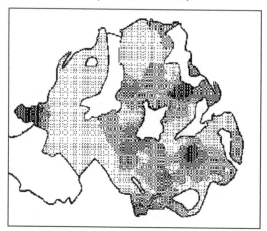

Figure 7.2: Incidence of mildew on spring barley in Northern Ireland in 1977, darker areas indicate greater incidence of disease; blank areas have no spring barley grown

predominant foliar disease of spring barley in Northern Ireland until recently (Mercer *et al.*, 1986), present from about growth stage 31 (Zadoks *et al.*, 1974) onwards. There is a range of cultivar resistance, but unlike *Septoria tritici*, some popular cultivars possess absolute resistance. This contributed to a marked drop in disease incidence in 1999 and 2000. Sowing resistant cultivars can also obviate the need for the application of fungicide sprays (Mercer and Ruddock, 1999). Attempts have been made by DARD over the past fifteen years to evaluate methods for the forecasting of these two diseases and these are described below.

Mildew on spring barley. The incidence of mildew is, to some extent governed by the weather during the growing season. Attempts were made by Polley and King (1973) and Polley and Smith (1973) in England to produce a forecasting model for mildew based on four weather factors. These were: (1) Day maximum temperature >15.6 °C; (2) Dewpoint depression at 09:00 >5 °C; (3) Daily rainfall <1 mm; and (4) Daily run of wind 246 km (integrated sum of wind speed over 24 h). The model indicated a day of high mildew risk if all four factors were satisfied. If three were satisfied on two consecutive days, that was also considered a high-risk day. The third consecutive day, on which at least two factors were satisfied, was also designated a high-risk day as long as one of the three days also had three factors satisfied. As soon as a high-risk day was identified, it was considered as the start of a high-risk period that would be terminated by a day on which no factors or only one was satisfied or the third consecutive day on which only two factors were satisfied. The model was tested in Northern Ireland with historical weather data over the period 1956-1979. The percentage of predicted mildew-risk days per month of the growing season

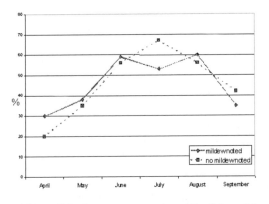

Figure 7.3: Average percentage of mildew risk days per month defined by Polley criteria for 1956-1979 for the years in which mildew was noted

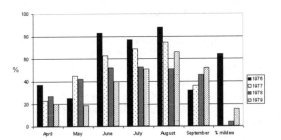

Figure 7.4: Percentage of mildew risk days per month in Northern Ireland growing season between 1976-1979 (growth stages 59-71)

was compared with the overall assessment of mildew incidence as indicated in the Annual Research Reports of the Department of Agriculture (Anonymous, 1957-1980). There was little difference in results for those years in which mildew was noted compared with those years in which no disease was noted (Figure 7.3). For the last four years of the exercise (1976-1979), average mildew incidence on the top three leaves of spring barley crops was assessed between growth stages 59 and 71 as part of a disease survey, and was compared with the percentage of mildew-risk days. Results (Figure 7.4) showed very good correspondence in 1976, a year with a high incidence of mildew, and a high percentage of mildew-risk days. There was also good correspondence in 1978, a year with a low incidence of mildew and a low percentage of mildew-risk days. However, 1977 had a considerably greater percentage of mildew-risk days than 1978, and yet had a very low incidence of mildew. In 1979, there was a moderate incidence of mildew, but a very low percentage of mildew-risk days, particularly at the beginning of the season. Correlation analysis of the mean number of mildew-risk days with % mildew gave an r^2 value of 0.64.

Control of mildew on susceptible spring barley cultivars appears to be very dependent on the time of spraying. A number of trials by Mercer and McGimpsey (1985) showed large differences in incidence of disease and subsequent increases in yield following a single spray of the fungicide propiconazole on the universally mildew-susceptible variety Golden Promise (Figure 7.5).

A series of experiments on two sites, Loughgall in Co. Armagh and Hillsborough in Co. Down, was set up in the years 1982-1984 to compare the effects of dates of

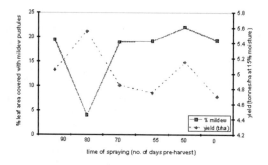

time of spraying (no. of days pre-harvest)

Figure 7.5: The effect of timing of single spray propoconazole fungicide on yield and incidence of mildew for Golden Promise spring barley (Northern Ireland, 1984)

application, predicted either from meteorological data or disease thresholds, of a single spray of the fungicide propiconazole on yield. The meteorological data were entered into the Polley model described previously, and the "best spray date" was calculated as the second day of a run of two or more days of mildew risk after 1 May. The date on which a disease threshold of 3% of the older leaves covered with mildew pustules was also noted. In 1982, the sowing date at Hillsborough was 19 April and at Loughgall, 2 April. In 1983 and 1984, both trials were sown on the same date, 14 April and 9 April respectively. The dates leading to maximum yield were then compared with predicted dates.

Results (Table 7.9) showed exact correspondence between dates predicted from threshold data and dates associated with maximum yield in 1984. The Polley method indicated a week earlier. In 1983, there was over a week's difference between dates for maximum yield between sites, but again the threshold method was closer in its prediction. In 1982, the difference between sites for dates for maximum yield was even greater at 16 days (probably partially reflecting differences in sowing dates) but the threshold and Polley methods were similar to each other and approximately at the median of the two site dates.

Table 7.9: Comparison of predicted "best" dates for single spray of propiconazole fungicide compared with spray dates associated with maximum yield for two sites in Northern Ireland from 1982 – 84

Year	Site	Predicted dates Polley method	Date of spray leading to best yield		
			3% mildew threshold*	Date	Days before harvest
1982	Hillsborough			17 June	77
1982	Loughgall	11 June	9 June	1 June	80
1983	Hillsborough			21 June	78
1983	Loughgall	30 June	22 June	13 June	69
1984	Hillsborough			1 June	80
1984	Loughgall	24 May	1 June	1 June	73

* mildew threshold data available only from Loughgall

Although there is an effect of weather on mildew, and although the Polley model did give reasonable correlation in some years, in other years the correlation was poor. This was probably at least partially due to not having local weather recording stations and the fact that the Polley model took no account of the growth stage of the crop. A better understanding of the effect of weather on mildew epidemics is also required. Until that is achieved, the use of a simple 3% disease threshold probably offers an effective option. This corresponds both to the point at which the disease is first readily seen when standing above the crop (Anonymous, 1985) and also to the start of the exponential phase of the disease epidemic.

Septoria tritici *on winter wheat.* Forecasting of a disease only makes sense if it is irregular in its severity and occurrence, and that at least some of this is related to weather conditions, otherwise routine control measures can be used. For example, the severity of *Septoria* diseases on winter wheat is dependent on rainfall, and thus more emphasis has been given to forecasting *Septoria* in England, where, because of drier weather, it is more irregular in its appearance and incidence than in Ireland.

An early attempt to produce a forecasting system was made by King (1972) with the following criteria:
⇒ Rain on two out of three days totalling at least 10 mm with rain on the first day
⇒ Rain on three consecutive days totalling at least 5 mm
⇒ Rain on four consecutive days
⇒ A relative humidity of at least 90% must be recorded at 09:00 GMT on at least one day in each period
    or
⇒ A period of sixteen hours or more commencing with rain and continuing or with a relative humidity of at least 90% with rainfall totalling at least 5 mm.

Tyldesley and Thompson (1980) discovered a correlation between disease incidence at growth stage (GS) 75 and number of days with at least 1 mm of rain between GS 37 and GS 59. The Agricultural Development and Advisory Service (ADAS) in Britain produced a similar model (Anonymous, 1984) of at least 1 mm of rain on any 4 days in the previous 14 days between GS 32 and GS 71. This model was subsequently modified to take account of 'splashy' rain events - ' a total of 10 mm or more in up to 3 consecutive days, once the canopy has reached full height, though over 5 mm on any one day may be sufficient in shorter crops where stem elongation is incomplete' (Thomas *et al.*, 1989)

Neither of the above models was tested in Northern Ireland, where the growing of winter wheat has only become popular in recent years. However, in 1988, DARD joined with Long Ashton Research Station, Rothamsted Research Station and ADAS in a study supported by the Home Grown Cereals Authority to test a model for *Septoria* forecasting developed at Long Ashton, based *inter alia* on the observation that the force of rain as well as its duration was important, because heavy "splashy" rain will spread spores more effectively than light rain, even of long duration (Parker *et al.*, 1994). A so-called "Splashmeter" was designed which consisted of a central pole, attached to which was a roll of receptor paper. Surrounding the pole, was a circle of small dishes containing dye that splashed up and marked the paper if heavy rain occurred. An empirical decision was made that only when spots were recorded above a certain height on the paper would this be considered as a "Splash-event". This information, along with a count of spore numbers on tillers at GS 30, allowed for a forecasting protocol to be set up (Figure 7.6), which forecast a maximum of one spray in a season with low initial inoculum, and a maximum of two sprays in a season with high initial inoculum. Forecast programmes were compared with spraying at fixed growth stages, and using both fixed growth stages and a forecast element. In Great Britain, highest yields (Table 7.10) were obtained with a two-spray programme at GS 37 and 59, a three-spray programme at GS 31, GS 39 and GS 59 or with the Long Ashton forecast system. The top yields, both of 8.05 t ha^{-1}, being with the latter programmes. In this instance, the forecast spray would

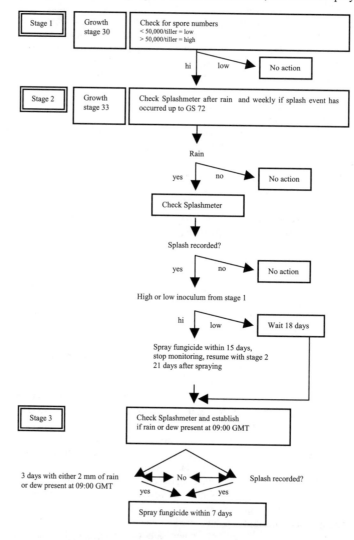

Figure 7.6: Scheme for forecast-guided control of Septoria

have been more economic than the three-spray programme, because only two sprays were applied. However, the difference in yield between the Long Ashton forecast programme and the fixed two-spray programme (traditionally the most effective programme if eyespot is not a problem) was insignificant at 0.06 t ha^{-1}.

In Northern Ireland, two trials, one on a first wheat and one on a second wheat, were set up in each of the three seasons 1988-1990, and compared a range of spray programmes with forecast ones. Unfortunately, due to interpretation problems, the fully forecast (Long Ashton) system was only applied correctly in 1990 when the second wheat was badly damaged by the root-infecting disease, take-all. However, in that year, the yield of the first wheat with the Long Ashton programme (Table 7.11) was comparable with the three-spray programme at GS 31, GS 39 and GS 59 and the two-spray programme at GS 37 and GS 59, which generally gave the most consistent yield return over the three seasons. However, overall there was considerable variation, a single spray at GS 31 giving one of the highest yields on the first wheat in 1988 and on the second wheat in 1989.

Table 7.10: Effect of a range of fungicide-spraying programmes on average yield of first and second winter wheats at Long Ashton and Rothamsted Research Station from 1988-1989

Programme	Long Ashton				Rothamsted		Mean
	1st 1988	2nd 1988	1st 1989	2nd 1989	1st 1988	1st 1989	
Untreated	7.92	7.28	6.74	6.80	7.66	6.43	7.14
ADAS forecast	8.34	7.24	7.23	7.05	8.49	7.18	7.59
LARS forecast	9.23	8.55	7.40	7.03	9.17	6.91	8.05
GS 31	8.80	7.84	7.07	7.06	8.51	*	6.55
GS 31 + forecast	9.20	8.32	7.32	7.08	8.44	7.00	7.89
GS 31 + GS 59	8.53	8.62	7.21	6.97	9.11	6.75	7.87
GS 31, 39 + 59	8.92	8.91	7.26	7.11	9.07	7.00	8.05
GS 37	8.50	7.83	6.91	7.14	8.30	*	6.45
GS 37 + forecast	8.71	8.28	7.30	7.06	8.35	7.00	7.78
GS 37 + GS 59	9.17	8.65	6.94	7.26	8.91	7.02	7.99

Table 7.11: Effect of a range of fungicide spraying programmes on yield of first and second winter wheats in Northern Ireland from 1988 - 1990

Programme	1988		1989		1990		Mean
	1st	2nd	1st	2nd	1st	2nd [1]	
Untreated	6.36	6.43	5.17	5.53	8.16	5.35	6.17
Long　　　Ashton Forecast	*	*	*	*	10.39	4.93	6.77
GS 31	7.04	7.47	5.23	6.03	9.07	5.78	6.77
GS 31 + forecast	*	*	*	*	9.91	6.23	7.20
GS 31 + GS 59	6.64	8.17	5.16	5.18	9.22	6.39	6.79
GS 31, 39 + 59	7.20	8.40	6.05	5.99	10.60	5.71	7.33
GS 37	6.84	8.05	4.68	5.28	9.59	7.18	6.94
GS 37 + forecast	*	*	5.79	5.98	10.31	7.12	7.16
GS 37 + GS 59	7.11	7.93	6.43	5.98	10.45	7.12	7.50

[1]trial severely damaged by take-all fungus

Although there is clearly a relationship between weather and the incidence and severity of *Septoria tritici* on winter wheat in Northern Ireland, the relationship is not particularly well understood. Current intensive testing of fungicide rates and timings alongside data on meteorological variables may help to clarify the relationship.

Decision Support Systems in cereal disease forecasting in Ireland
In the past few years, various research workers (Secher and Bouma, 1996; Audsley *et al.*, 1997; Paveley *et al.*, 1997) have begun investigating the use of DSSs which, it is hoped, will be able to take into account the large number of factors, including past, present and future weather conditions, which need to be considered in the growing of cereal crops. One such DSS is the computer-based programme PC-Plant Protection developed by the Danish Institute of Agricultural Sciences, Lyngby, Denmark (Secher and Bouma, 1996). The programme incorporates models for pests, diseases and weeds. In particular it gives risk indices for aphids, powdery mildew, *Septoria* and yellow rust. The risk indices use local weather data obtained from a Hardi Metpole automatic weather station (Hardi International) to estimate the development of pests and diseases. The weather data are the number of days with precipitation >1 mm in periods defined by the programme; five day weather forecasts; number of days with precipitation >1 mm; daily maximum temperature; average temperature and precipitation in April.

At Oak Park Research Centre, Carlow, Ireland, trials using risk indices for *Septoria* control obtained from this programme are being carried out. The trials started in 1998 and will continue for a number of years. Two methods are being used to identify the time of spraying: (1) a leaf diagnostic test for the presence of *Septoria* spp. and (2) the PC-Plant Protection risk index. Fungicides are applied based on these two prediction systems, and are compared with two standard fungicide programmes and an unsprayed treatment. One standard fungicide programme consisted of 0.9 l ha^{-1} Sportak (prochloraz 450 g active ingredient l^{-1}) at GS 31/32, a tank mixture of 0.3 l ha^{-1} Opus (epoxiconazole 125 g active ingredient ha^{-1} and 0.7 l ha^{-1} Amistar (azoxystrobin 250 g active ingredient l^{-1}) at GS 39 and 0.8 l ha^{-1} Amistar at GS 59. The second fungicide programme was a two-spray programme with the GS 31 application omitted.

Results for the trial in 1998 are shown in Table 7.12. All spray programmes yielded significantly higher than the unsprayed controls. There was no yield benefit from applying fungicides in response to latent infections (diagnostic test) or to PC Plant Protection risk indices when compared to routine applications. However the timing of the second spray varied with the system used. In the routine application at GS 39, the date of this spray was 18 May, the diagnostic test recommended 28 May as the spray date while the DSS risk index indicated 13 May. When the GS 31 application was omitted, the spraying dates for the second spray remained the same for the routine application and for the PC plant protection risk index but the diagnostic test recommended a spray date ten days earlier (18 May).

The benefits from the use of DSS aids to spraying in an Irish context may be useful on two counts: (1) to indicate the most appropriate spray timing and (2) to reduce the number of sprays applied, e.g. in a low disease year or when more resistant cultivars to *Septoria* become available. However, although the use of DSS offers a chance to integrate the many variables that need to be considered in formulating a fungicide-spraying programme, the task is a formidable one. This is especially so in Northern Ireland, where the relatively high rainfall frequently restricts the ability of the farmer to get on the land and spray at the optimum time. For DSSs to be adopted they will have to be seen to improve performance above that currently achieved, generally with fixed growth stage sprays.

In Northern Ireland, DESSAC (Decision Support System for Arable Crops) sponsored by the Home Grown Cereals Authority is currently being evaluated for winter wheat, but results are not available yet (2001).

Table 7.12: Effect of fungicide programmes on disease control and yield response in winter wheat *cv.* Brigadier in 1998

Treatment	Timing	Application Date	Yield t/ha @ 15%	% Septoria 2nd Leaf (GS 75)
Sportak	G.S. 31	15 April		
Opus + Amistar	G.S. 39	18 May		
	G.S. 59	11 June	7.46	38
Sportak	G.S. 31	15 April		
Opus + Amistar	Diagnostic leaf 3	28 May		
	Diagnostic leaf 2	19 June	7.37	41
Sportak	G.S. 31	15 April		
Opus + Amistar	DSS	13 May		
	DSS	11 June	7.42	53
Opus + Amistar	G.S. 39	18 May		
	G.S.59	11 June	7.04	59
Opus + Amistar	Diagnostic leaf 3	18 May		
	Diagnostic leaf 2	11 June	6.98	60
Opus + Amistar	DSS	13 May		
	DSS	11 June	7.26	60
Unsprayed			4.59	100
L.S.D			1.02	15

7.2.3   Crop pests

Cereal Aphid Populations

Cereal aphids cause damage directly to crops by sap-sucking, but also indirectly by the transfer of virus particles such as barley yellow dwarf virus (BYDV). Spread, growth rate and reproductive rate of the aphids is at least partially dependent on weather factors, such as temperature (Acrema and Dixon, 1989) and rainfall (A'Brook, 1981). The most important species of grain aphid to attack cereals is *Sitobion avenae*. The summer population on cereals develops from winged migrants (alates) or resident overwintering aphids. Watson and Carter (1983) derived a relationship for the number of *Sitobion avenae* alates caught before flowering (GS 69) in a suction trap in East Anglia in England and the accumulated cold degree-days (below 0°C) over the winter months. The size of the migration depended on the production and survival of alates over the winter - mild winters favoured high alate numbers and *vice versa*. However, work in Northern Ireland (A. Bell and P. Mercer, unpublished) showed only a poor correlation (r = - 0.48) between numbers of *Sitobion avenae* caught in a suction trap in April and numbers of cold degree-days from October to February over the period 1976-1984. In Ireland, similar work (T. Kennedy, unpublished) showed no correlation between numbers observed in winter barley crops in April and numbers of cold degree-days from October to February over the period 1989-1998. Part of the explanation for the difference between regions may lie in the relatively mild climate of Ireland compared to that of south-east England.

While the size of the spring migration of aphids before GS 69 is related to over-winter survival, its timing has been found to be dependent on temperature during the early months of the year. Watson and Carter (1983) showed that warmer than normal temperatures in January and February resulted in earlier migration. However, Vickerman (1977) also showed that a cold March and April preceded delayed but significant aphid outbreaks in summer, both in terms of peak numbers and the number of 'aphid days'. It was postulated that cold springs are associated with low levels of natural enemies to prevent the rapid build up of aphid populations. In contrast, when aphid populations on grasslands and primary hosts were high in spring, there was a rapid build up in populations of predators and beneficial insects, and aphid outbreaks did not occur in summer. In a study in south-east Scotland, Sparrow (1974) derived relationships between the dates of peak aphid arrival after May 1 and deviations from the mean temperature in March. For *Sitobion avenae* the dates range from May 1 for a deviation of plus 1 °C to May 1 plus 70 days for a deviation of –2 °C whereas for *Rhopalosiphum padi* the dates ranged from plus 10 to plus 37 days respectively.

A'Brook (1981), in Wales, showed correlations of numbers of different aphid species in June and July with combinations of rainfall and day-degrees above and below 6 °C, e.g. high numbers of *Rhopalosiphum padi* in September and October were correlated with high numbers of day degrees above 6 °C in those months. Research in Northern Ireland (A. Bell and P. Mercer, unpublished) showed a similar correlation using numbers of aphids captured in a high level suction trap from 1976-1984 and weather data obtained from the synoptic weather station at Aldergrove, Co. Antrim. However, the researchers were not able to confirm the positive correlation between number of *Rhopalosiphum padi* in September and October with summer rainfall. They did find, on the other hand, a significant regression predicting numbers of *Rhopalosiphum padi* in October from rainfall between March and June (r = 0.94):

$$\log (n+1) = 0.008 \, DD - 0.007 \, R + 2.0 \qquad \text{7.1a}$$

where n = total number of aphids caught in October; DD = accumulated day-degrees above 6 °C in September-October; and R = rainfall (mm) in March-June. This finding was confirmed by results from Ireland (T. Kennedy, unpublished) (r = 0.90):

$$\log (n+1) = 0.0025 \, DD - 0.0097 \, R + 2.23 \qquad \text{7.1b}$$

where n = total no. of aphids observed in winter barley fields from October to December, DD = accumulated day-degrees above 6 °C in September-October and R = rainfall (mm) in March-June. This relationship may be explained by spells of heavy rain in the spring/early summer period leading to an increase in mortality, particularly during migration and colonisation in May when aphids are not protected by the ear and are thereby vulnerable.

Monitoring and forecasting aphid numbers is only part of the picture. Acquisition and multiplication of the viruses which are carried by the aphids are also affected by weather (Lowles *et al.*, 1996). Further investigation is necessary to test and extend models to provide more precise predictions of spring and autumn migrations of the various aphid species, virus infectivity and predator activity levels. In the meantime, greater awareness, using the above criteria, of the factors that determine aphid numbers will facilitate better decisions for protective measures.

7.3   Prediction of Weather-related Animal Diseases and Stress

Introduction

In temperate, moist climates, animal health management is often concerned with parasitic diseases, which cause a lack of thrive or productive conversion in animals, much loss of milk, reduced reproductive efficiency, and at times the death of the host animal. For seasonal peaks of disease, or in epidemic situations, models predicting the effect of weather on the development of various parasitic organisms can provide valuable objective guidance to the advisory services. Preventative measures against epizootic viral diseases, such as foot-and-mouth disease, also benefit from specific guidance based on meteorological data so that effective measures can be put in place to stamp out an outbreak before it becomes endemic in the national herd.

An understanding of the critical temperature-related stresses on animals, which occur during harsh weather conditions, is equally necessary for effective stock management. Certain diseases, e.g. pregnancy toxaemia, grass tetany and swayback, are associated with cold weather, while stress associated with excessive heat also weakens the immune system. Although sheep are well adapted to cold conditions, e.g. hill breed ewes successfully raising lambs on a hill or mountainside, very often sudden changes caused by the combined effects of cold, rain and wind, not necessarily severe, can be very significant. Wind can open the fleece and driving rain can enter into it. The new born lamb, particularly less than 48 hours old, is especially vulnerable to hypothermia and high mortality can occur during harsh conditions. In general, relatively high losses (sometimes 10-20 per cent of all deaths) can occur from cold stress in sheep and lambs (Duncan, 1998; O'Brien, 1989; Starr, 1988). A number of models have been developed to simulate and calculate the climatic factors which influence animal parasitic life cycles, trigger stress in cattle and sheep and contribute to the spread of viral diseases. The chemical effects of some parasitic disease outbreaks are the accumulated result of the meteorological conditions prevailing over several months whereas other diseases and stresses stem from short-term daily weather conditions. The examples given in the following sections exemplify the possibilities for the application of a range of models to different diseases in temperate climates. Other models may be found in the appropriate literature (Wall *et al.*, 1993; Anonymous, 1989; Starr, 1988; MacCarthy and Lynch, 1986; Gibson, 1978), some of which may need updating and local validation.

7.3.1 Fascioliasis

Ollerenshaw Model

Fascioliasis is a disease of the liver and is caused by the liver fluke, *Fasciola hepatica*. This is an example of a parasitic disease with two hosts, a direct host such as cattle or sheep, where the sexual stage occurs, and an indirect host, a mud snail (Lymnaea truncatula), where the asexual stage takes place. The ability of the parasite to complete the stage of its life cycle outside the host animal is highly weather dependent. As the critical developmental conditions relate to recent weather rather than future weather, the likelihood of the next stage of the cycle taking place is predictable; hence the disease is preventable and treatable. Fascioliasis is a disease of warm and wet conditions. Weather affects the parasite in many ways. If the temperature is low the fluke's life cycle will not actively progress, nor indeed will that of the mud snail. Egg development within the snail is dependent on the ambient temperature and is halted below an air temperature of $10\,^0C$; continuous immersion in water or mud is needed for development of the snail and wet conditions are required for the emergence of young fluke from the snail. The Ollerenshaw Index (M) (Ollerenshaw, 1966; Thomas, 1978) is widely used in Ireland as a predictive model to evaluate the climatic conditions of a season for liver fluke development. The model uses monthly estimates of potential evapotranspiration (ET_p, mm) and measured rainfall totals (P, mm), together with the number (N) of raindays (days with 0.2 mm or more), to derive the monthly contribution (M_t) to the season's index in the following manner:

$$M_t = [(P - ET_p)/25.4 + 5] \cdot N \qquad\qquad 7.2$$

Where the monthly total, M_t, exceeds 100 (as in a wet month) it is given a maximum value of 100. The constant 5 is used to ensure M_t has a positive value and the value 25.4 arises because the original formula was derived when rainfall was measured in inches. Since the development stage of the snail is slower in May and October due to lower temperature, the maximum value allowed for M_t in those months is halved, i.e. 50. Summed over May to October, the total M, commonly called the Ollerenshaw 'Summer Index' to indicate summer infection, has a maximum possible value of 500. Actual M values are assessed as follows:

M ≤ 320	little or no disease expected
M > 320	occasional losses may occur
M > 450	disease may be prevalent and acute
M > 480	potentially epidemic situation

From the life-cycle it is clear that the effects of summer infestation picked up by grazing livestock become apparent in the exposed animals over the following winter, hence the index is perceived as a predictive one.

The index of severity of infestation arising from the parasitic larvae overwintering in the snail is estimated using a similar calculation for the months of August to October of one year and May and June of the following year. With limits to M_t for the two months, October and May, set at 50, and summation made over five instead of six months, the possible maximum value of the overwinter index is 400. The following criteria for expected severity of infection are suggested:

M ≤ 250	little or no infection expected
M > 250	occasional losses may occur
M > 320	disease may be prevalent and acute
M > 380	potentially epidemic situation

Figure 7.7 shows the Ollerenshaw overwinter (1998/99) and summer (1999) indices respectively at synoptic stations (see Chapter 3), with interpretation of the data. Considerable regional differences in the levels of fluke infection in animals following both seasons are predicted from the data. Figure7.7(a) for overwinter indices indicates that, except in the south-east of the country, infection in sheep and cattle is likely to be prevalent (with the potential for epidemic conditions to occur on higher ground) over the succeeding summer and autumn months. From the summer indices (Figure7.7(b)), infection leading to severe fascioliasis in animals can be expected over the winter of 1999/00 in the north-west but only occasional losses due to fluke are likely to occur in the south-east. Veterinary Authority advice to farmers on animal dosing strategy during the succeeding months takes these considerations into account (although the Ollerenshaw indices have not been expressly validated under Irish conditions, they have been found to be very useful).

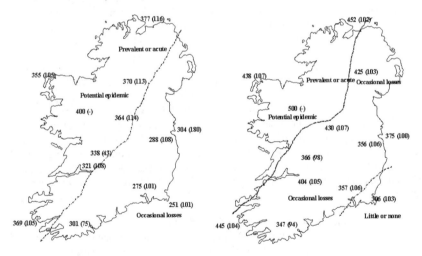

Figure 7.7: The (a) overwinter (1998/99) and (b) summer (1999) Ollerenshaw indices at synoptic stations (with per cent of the long-term average in brackets)

The Multivariate Model
A liver-fluke predictive model based on meteorological data alone cannot provide accurate guidance. Even a multivariate weather model does not significantly improve on the Ollerenshaw method (McIlroy *et al.*, 1990). Established levels of acute fatal

fascioliasis in sheep do not necessarily indicate future prevalence, however a time series analysis of Northern Ireland data showed that the previous year's prevalence of fascioliasis accounted for 67 per cent of the variation recorded. Thus McIlroy *et al.* (1990) considered the effect of the previous year's severity of infection together with the current year's weather variables on the subsequent prevalence of the disease in a multivariate formula. Five weather variables were included in the model and regressed against the time series of the residual levels of fascioliasis. The equation for the weather effect, W_t, has the form:

$$W_t = 38.4 - 5.0W_1 + 1.6W_2 - 2.5W_3 + 1.0W_4 + 1.8W_5 \qquad\qquad 7.3$$

where W_1, W_2 and W_3 are the summer (June, July, August) mean wind speed (m s^{-1}, but originally in knots), mean (average monthly of three months) rainfall (mm) and mean air temperature (^{0}C) respectively; W_4 and W_5 are the spring (March, April, May) mean wind speed (m s^{-1}) and mean air temperature (^{0}C) respectively. It was observed, however, that animal management factors also played an important role in the level of fascioliasis encountered. With the inclusion from 1983 onwards of the observed percentage perturbation in disease levels of the previous year as a categorical variable, the authors were able to quantify more accurately the potential risk in any year (accounting for 95% of the variation).

Data on the annual percentages of sheep livers with fascioliasis at slaughter in Northern Ireland, extracted from the McIlroy *et al.* (1980) study, were re-analysed in conjunction with the corresponding meteorological data from the nearby weather station at Clones, Co. Monaghan. By a method of least squares regression and successive approximation the following formula was derived:

$$D \approx 9.2 + 0.9\Delta Z_{t-1} + 0.04W_t \; (r^2 = 0.93) \qquad\qquad 7.4$$

where D is the forecast percentage of sheep livers with fascioliasis which are condemned at abattoirs for the upcoming year, ΔZ_{t-1}, represents an adjustment for recent (past year) changes in management practices, and W_t is the current year's weather parameter given by equation 7.3 above. ΔZ_{t-1} can be derived using equation 7.3 on the verifying data (D and W_t) of the immediate past year. If management is unchanged, i.e. ΔZ_{t-1} is zero, and weather conditions are average, i.e. W_t equals 100, then the formula gives D for sheep as 13.2 per cent.

The formula was tested on data available for 1996-1999. In 1996, W_t was 139 and the observed annual (August to July, 1996/97) national sheep liver condemnations was 18.6 per cent. Thus a perturbation (ΔZ_{t-1}) of a 3.8% deterioration in management from the long-term average, is derived. With W_t as 116 in 1997 and the incoming ΔZ_{t-1} of 3.8 per cent, an infection level of 17.3 per cent was predicted for 1997/98, which compared reasonably well with abattoir condemnation rate of 15.4 per cent. From an extreme weather parameter of 198 in 1998, and a ΔZ_{t-1} (the residual in 1997/98) of 1.6, well above average infection level at 18.6 per cent was indicated for 1998/99; this compared favourably with observed condemnations of 16.9 per cent. If liver condemnations at individual meat plants or abattoirs were used together with weather data from the nearest meteorological station, then more accurate regional predictions may be possible. Also more work needs to be done at abattoirs to separate out livers condemned due to liver fluke infestation and those damaged due to other causes.

Temperature Model
The work carried out by McIlroy *et al.* in the mid-1980s has now been superseded to some extent by new work in this area by Catney (1997). From this recent work, the most efficient model for forecasting the level of condemnations due to fascioliasis in sheep was found to be:

$$Y = Y_{t-1} - 2.33 \ (T_1 - 13.8) \qquad\qquad 7.5$$

where Y is the forecast level of fascioliasis (between August and the subsequent July, i.e. the fascioliasis year), Y_{t-1} is the level in the fascioliasis year immediately prior to the forecasted year, T_1 is the mean air temperature (^{0}C) over June, July, August for the year Y_{t-1}. The model, which is based on data collected at all meat plants in Northern Ireland, would clearly need some tuning before being used elsewhere. This type of model corresponds closely with similar work on liver condemnations in cattle reported by Goodall and Menzies (1995).

Extra-Mammalian Prediction Algorithm
Practical models for predicting the temperature-dependent survival and time required to complete the development of the extra-mammalian stages of the liver fluke, *Fasciola hepatica,* under varying temperature conditions in temperate moist climates (e.g. Ireland) were formulated and tested by Hope Cawdery *et al.* (1978). These models were then combined with certain biological parameters to make a prevalence-prediction algorithm. A prediction algorithm using temperature and moisture models was developed, but needs further validation and refinement before it can be put into routine use.

7.3.2   Nematodiasis

Nematodiasis is a parasitic disease which, in particular, adversely affects the health and growth of new season grazing lambs. The eggs of one season do not hatch during that season, but overwinter and hatch the following spring at the time that new lambs are starting to graze. The development stages of endoparasites on grass require moisture and an ambient (air) temperature above 10 ^{0}C. One parasite, *Nematodirus immitis,* which affects young lambs (older animals are resistant) not only requires the appropriate weather conditions to thrive, but also preconditioning by a period of cold (below 4 ^{0}C). While the spring rise in temperature is important to initiate larval hatching, the date of greatest exposure of lambs on grass, (i.e. period of highest risk), can be reasonably forecast, based on temperature during March (Thomas, 1978). The month is subdivided into overlapping periods, e.g. March 1 to March 15, March 1 to March 20, March 1 to March 25 and March 1 to March 31, to allow estimates to be made earlier. As each period is completed the average soil temperature (say at 100 mm depth) is obtained for that period and used in a similar type formula, but with different parameters. Given the limits of expected accuracy, and as the four formulae by Thomas (1978) provided quite similar results, a compromise formula has been adopted, as follows, for the prediction of the number of days, N, after March 31 when maximum larvae are available on the herbage for ingestion by the young lambs. This is given as:

$$N \approx 53.3 - 7.1\ T \qquad\qquad 7.6$$

where T is soil temperature in ^{0}C. If the mean soil temperatures for the first 15, 20 and 25 days of March are 6.2 ^{0}C, 6.4 ^{0}C and 6.6 ^{0}C respectively, then the corresponding predicted dates after March 31 are April 9, April 8 and April 6. Thus the dates become refined as time progresses. A delay in hatching is said to be associated with dry periods. The author suggests that 1.5 days be added to the observed (predicted) date of pasture peak for each dry day in excess of 25 from March 1. In moist, temperate climates this correction is rarely necessary.

7.3.3   Cold Stress in Animals

Wind Chill in Sheep

Wind chill in sheep can be simulated by the calculation of sensible (non-evaporative) heat loss (H_n) that sheep experience as a result of exposure to the weather variables of air temperature, wind speed, sunshine, cloud and rain (Mount and Brown, 1982; 1983). The effect of wind chill can be evaluated for periods when heat loss exceeds 55 Watts (W) m^{-2}, the rate that is expected at the critical ambient temperature. The following formula provides an estimate of the heat loss from sheep:

$$H_n = \frac{T_{bn} - T_a + I_a[K_R(10 - 9c) - \alpha R_a]}{I_t + I_f + I_a} \qquad\qquad 7.7$$

where:

H_n heat loss in W m^{-2}

T_{bn} deep-body temperature for sheep (37 ^{0}C)

T_a air temperature (^{0}C)

I_t tissue insulation (~ 0.12 K m^2 W^{-1})

I_f fleece insulation given by: $f(1 - 0.3t_w)/(90 + 14\ V_a)$ (K m^2 W^{-1}) with f as fleece depth in mm (taken as 30, 50 or 70 mm, as appropriate), t_w is proportion of time with a wet fleece and V_a is wind speed at sheep level, m s^{-1}

I_a surface insulation given by: $(1 - 0.3t_w)/(9 + 4\ V_a)$ (K m^2 W^{-1})

K_R radiant heat transfer coefficient for long-wave radiation exchange (5.6 K m^2 W^{-1})

c fraction of sky covered by cloud

α absorption coefficient for solar radiation (0.75 for sheep fleece)

R_a solar radiation, Wm^{-2}, incident on the sheep.

Wind speed at sheep level is taken as 0.4 of the 10-metre height wind speed. Mount and Brown (1982) also recommended using hourly rainfall rate to quantify t_w. They proposed a value for t_w of 1 for ≥ 1 mm h^{-1} and a fraction of 1 for < 1 mm h^{-1}, i.e. for 0.3 mm h^{-1}, $t_w = 0.3$. Calculation of the quantities is made using hourly values of air temperature, wind speed, sunshine, cloud and rain. H_n is given as a daily mean of the hourly results and $H_n = 55$ Wm^{-2} is taken as the sensible heat loss from a grazing sheep at the animal's critical temperature. Below the 55 Wm^{-2} threshold, the heat loss from the animal is compensated for by metabolic heat production. It is thus useful to compare the heat loss with this quantity. It is also useful to derive the heat loss for wind speed alone by recalculating the above formula for H_o with V_a set to zero, and thus H_v, the component of heat loss that is due to wind, from $H_v = H_n - H_o$.

Pneumonia in Sheep
A close correlation between the percentage lung condemnations for pleurisy and pneumonia in sheep and adverse weather prevailing during the same month, the previous one and previous two months in Northern Ireland, was reported by McIlroy *et al.* (1989). The most significant correlation was found with wind speed. A rain/windchill factor (R_W) was calculated, taking into account four weather variables as follows:

$$R_W = WH(35-T) + 50R \qquad 7.8$$

where: W, is the wind speed (knots; 1 knot = 0.51 m s^{-1}); H, air relative humidity (%); T, air temperature (^{0}C); and R, rainfall (mm). The equations obtained by regressing percentage condemnations (y) and the above rain/windchill factor, measured in the same month, the preceding month and two months previously, are given by:

$$y = 0.131 + 0.00105 R_W$$
$$y = 0.0115 + 0.00147 R_{W-1}$$
$$y = 0.0108 + 0.00147 R_{W-2} \qquad 7.9$$

The standard errors (SE) of the regression were 0.0002, 0.0004 and 0.0002 respectively. Warnings could usefully be issued when the meteorological conditions are such that the index exceeds a critical value. An index of 350, which typically corresponded to about 20 per cent prevalence of pneumonic lesions in sheep slaughtered, might provisionally be accepted as a critical threshold.

Wind Chill in New-born Lambs
New-born lambs have low body energy reserves and are highly vulnerable to cold stress. Mortality due to wind chill during and soon after birth can sometimes account for significant death losses in lambs. Cold, wet conditions can readily induce heat loss in a new-born lamb (up to 48 hours old) exceeding its 'summit metabolism' – the maximum rate of heat production that the lamb can sustain. Based on farm data near Reading, England, and meteorological variables such as wind speed, air temperature and duration of rainfall, and applying the results of published research, Starr (1981, 1984) devised criteria for operational forecasting the wind chill factor for young lowland lambs. The occurrence of rain is crucial for most stressful conditions, so a contingency table (Table 7.13) is provided giving separate indices representing wet coat and dry coat situations.

Table 7.13: Wind-chill indices for wet and dry coat new-born lambs*

Wind speed	Air temperature (^{0}C)						
	Wet coat indices				Dry coat indices		
(ms^{-1})	> 5	5 to 0	0 to –5	< -5	> 0	0 to –5	< -5
> 10	17	18	19	20	12	13	15
5-10	16	17	18	19	11	12	13
< 5	14	15	16	17	10	11	12

*After Starr (1984).

As fluctuations of weather from hour to hour affect the lamb heat demands, the index is determined for each four-hour interval (from the mean four hour wind speed and temperature, and the occurrence or otherwise of rain); a sum of the 6 indices is arrived at for each day (M_{24}) for heat loss. The following thresholds are proposed: If $M_{24} > 90M_B$ (basal metabolic rate), stress conditions can be critical; between 80 and 90 M_B - danger; between 70 and 80 M_B - warning; and <70 M_B - little danger. Starr (1984) concluded that the scheme overestimated the wind-chill under daytime conditions of strong insolation (sunshine), but that it was realistic under severe night-time conditions. As multiple lambs may have to compete for the mother's milk supply, and if there is pressure on housing, a practical approach for turn out was adopted by one of the farmers participating in the study using M_{24} wind-chill criteria as follows:

Wind-chill factor
81 - 90:   Turn out only single lambs
71 - 80:   Turn out twin lambs
≤ 70:   Turn out triplet lambs

7.3.4   Foot-and-Mouth Disease

Foot-and–Mouth Disease (FMD) is a highly contagious disease of cloven hoofed animals, e.g. cattle, sheep, goats, pigs. While the movement of infected animals is the single most important means of spread of FMD (other means are slurry, urine, faeces and feeding of infected milk), secondary exposure of other animals is principally the result of airborne virus dispersion. The latter form of spread is related to low level winds during periods in which infected animals shed the virus. Veterinary authorities put strict controls in place as soon as an outbreak of the disease is reported so as to contain and stamp out the disease immediately, e.g. restriction in direct animal to animal contact, removal of contaminated food products, slaughter and disposal of all animals from infected farms. Nevertheless, the airborne route essentially remains an uncontrollable means of transmission of the disease.

The main factors which influence the airborne spread of FMD are the quantities and type of virus emitted from infected animals (primary sources), the period of emission, pattern of virus dispersion and the amount of virus available to susceptible animals (secondary outbreaks). Thus, the risk of spread of the disease depends on the species of animal emitting the virus, the stage of the disease and the number of animals affected. Infected pigs liberate virus at a rate of 277,000 $TCID_{50}$ (bovine Thyroid Tissue Culture Infectious Units) of airborne virus per minute, as against 170 $TCID_{50}$ per minute for steers or sheep. The incubation period is usually 2 to 14 days.

The emitted virus is encapsulated in a small respiratory moisture droplet exhaled by the infected animal and remains viable so long as the droplet remains intact. An atmospheric relative humidity (RH) greater than 55% has been found to permit the virus to survive. The droplet can be carried downwind from the source and dispersed by direct transport in the wind and by diffusion, or spreading sideways and vertically through the wind stream by turbulence. Although a simplification of reality, the following form of a Gaussian type model can be used to simulate the dispersion of an aerosol or pollutant from a release point source (Gloster et al., 1981):

$$C_{xy} = \frac{Q}{\pi U_{10}\sigma_y\sigma_z} e^{\frac{-y^2}{2\sigma_z^2}}$$ 7.10

where C_{xy} is the concentration at the co-ordinate (x,y), Q is source strength (taken in this case to be at ground level), U_{10} is wind speed measured at 10 m height, σ_y, σ_z are dispersion coefficients in the y and z directions. The plume diffuses across wind (y-direction) and in the vertical (z-direction). As the plume spreads downwind, concentrations within it decrease, becoming diluted in an increasing volume of air. The distribution of the concentration in the y and z-directions has standard deviations of σ_y and σ_z (Pasquill, 1974). Within these limits, the plume retains some 67 % of the concentration with maximum value along the central line. An idealised depiction of a Gaussian plume is given in Figure 7.8.

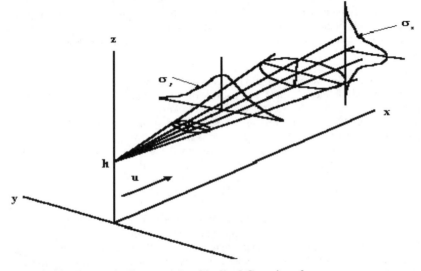

Figure 7.8: Schematic diagram of an idealised Guassian plume

The value of the dispersion coefficients, σ_y and σ_z, increases downwind at rates which depend on both the turbulence in the air and on the wind speed (and direction). For five of the stability classes, ranging from Class F for very stable air to Class A for very unstable turbulent air (see Pasquill, 1974), σ_y varies on a logarithmic scale from about 35 m to 215 m at 1 km, and 270 m and 1550 m respectively at 10 km distance from source. The corresponding variations in σ_z are 14 m and 450 m at 1 km, and 45 m and 1350 m (for class B, but infinitely large for class A) at 10 km. These values are very approximate, and are but an exemplification of the model values of the dispersion coefficients, which are usually derived by formula from the meteorological data.

Animals are more likely to be infected by inhalation of the virus than by ingestion. Thus, virus rained out of the air onto herbage is essentially not available as a source of secondary infection. The model incorporates precipitation by assuming none of the emission is available for infection away from the source for those hours for which precipitation is reported in the meteorological data. Another consequence of the inhalation route is that those animals with largest air intake, e.g. cows and steers as against pigs or sheep, have greatest potential for acquiring secondary infection.

To operate the FMD model the following information is required: an estimation of the duration and total quantity each day of airborne FMD virus dispersed from the infected premises or herd; hourly or 3-hourly observations of wind speed and direction, relative humidity, cloud cover and precipitation in the vicinity of the outbreak, and latitude and topographical features of the area. Atmospheric stability values are derived from cloud type and amount as well as surface flux parameterisations. In the early stages of an outbreak, and in the absence of information on source strength, default virus concentrations at source are used as inputs to the model. The meteorological data for the dispersion model are taken from a representative weather station nearest to the outbreak. Alternatively, data may be obtained from the outputs of numerical weather models (see Chapter 8), although due to some smoothing in numerical models, such data may not have sufficient accuracy unless a fine scale, and perhaps a non-hydrostatic, model is used.

The output data are processed at grid points radially extending out to 10 km from initial source spaced at intervals of 1 km and at 10 degrees intervals from 0-360° polar (radiating) co-ordinates. It is considered that the 10 km distance from source is sufficient limit for over 90% of the virus to have been deposited. Outside of this range the probability of secondary infection occurring is minimal. Topographic features are also included as inputs to provide greater realism to the outputs. A gradient of 1:50 is allowed in the model to deflect the wind. The outputs are given in terms of dose per day required to infect animals. High risk situations occur when cattle are present in large numbers and inhale infected air for a prolonged period.

The output data may be produced in the form of a table or plotted on a graph using polar co-ordinates centered at the point of emission. Such diagrams display features such as the interaction of local terrain and wind direction. Trials conducted by Pirbright Veterinary Laboratory (Donaldson *et al.*, 1987) have shown that an estimated virus dose of 25 $TCID_{50}$ per day was likely to infect cattle, 15 $TCID_{50}$ to infect calves, sheep or goats, and 400 $TCID_{50}$ to infect pigs. Donaldson *et al.* (1987) also found that the dose depended on the virus strain.

Computerised Support System – EpiMAN®
During an animal health emergency, time and resources need to be conserved as much as possible. Mistakes can occur when making numerous complex decisions under pressure. The continuing reduction in manpower in State veterinary services and the increased use of information technology (IT) has prompted interest in the use of management information systems. For the control of FMD, where the airborne virus challenge to animals at risk can be estimated and predicted using airstream and diffusion models, the incorporation of FMD models into a management information system is highly desirable.

A computerised support system to assist in control of outbreaks of FMD was developed in New Zealand (Sanson, 1994). Its feasibility for use in a European context was tested under an EU 3-year funded project as reported by Mackay *et al.* (1997). The EpiMAN® system consists of a series of models and expert systems (Figure 7.10). The software comprises a number of PC 'clients' linked to a central database and a knowledge base, which is managed by a Geographical Information System (GIS). The knowledge base consists of a number of mathematical models, predictive models and expert systems which contains information on the epidemiology of FMD. The results are stored in the database.

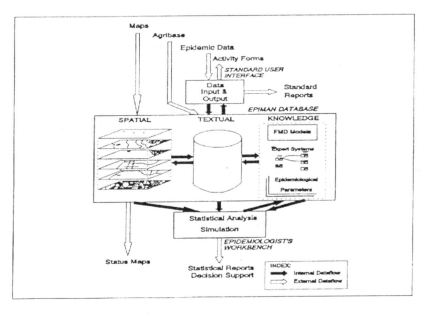

Figure 7.10: The Structure of the EpiMAN® computerised expert system (Anonymous, 1997)

There are four models of FMD contained within the EpiMAN® system. Two of these are the FMD virus production model (an on-farm infection model which quantifies virus release to the atmosphere) and the meteorological diffusion model (e.g. Gaussian dispersion or Gaussian puff (Sørensen and Christian, 1996)). The latter model uses the most recent weather conditions from an existing meteorological station, a specially erected on-farm weather station, or the output from a Numerical Weather Prediction model to predict diffusion of the virus. The third model is an inter-farm spread model that can simulate an entire epidemic or can pick up the state of the epidemic at any given time and simulate forward a user-definable time period. The fourth model is a simple deterministic model that estimates the dissemination rate throughout the epidemic and then extrapolates forward to new IPs (infected premises) and to the conclusion of the epidemic. The EpiMAN® DSS can also be adapted to other contagious diseases, e.g. swine fever or bovine tuberculosis.

7.4   General Conclusions

The use of weather-driven models in the simulation and prediction of animal diseases and stress has made slow but steady progress over several decades. There are now new ones for several other diseases, such as sheep fly myiasis (fly strike) developed by Wall *et al.* (1998). The number of models, capable of off-the-shelf transfer to other regions without local validation, however, is as yet limited. Out of concern for animal welfare and the regulation of preventative measures and appropriate treatments, and to avoid health stresses and productivity losses, greater precision in decision making is required. The representative models treated in this chapter are parasite, virus or stress related. While many of the models are quite general and give imprecise information, they do enable farmers and the veterinary authorities to have a greater awareness and understanding of the importance of climate on livestock performance by providing a strong basis for forecasting the impact of recent and current weather on outdoor animals.

Forecasting and model-making of plant and animal disease and health problems began in the early to middle part of the twentieth century. Many of the basic relationships discovered then are still at the heart of the more sophisticated models which are now being built round them. However, the relatively recent availability of powerful and cheap computing capabilities has enabled many more possible contributory factors to be taken into account in the initial development of forecasting models and also to handle the large quantities of field and meteorological data which are subsequently fed into them. The employment of Geographical Information Systems (see Chapters 2 and 6) based management systems, integrating disease models with various data sources and expert systems, is the key to maximising the value of the information provided by these weather driven models that are available. However, models must be user-friendly, robust and capable of giving better outcomes in terms of economics and environmental impact than existing prophylactic systems.

References

A'Brook, J. (1981): Some observations in west Wales on the relationships between numbers of alate aphids and weather. *Annals of applied Biology* **97**: 11 – 15

Acrema, S.J. and Dixon, A.F.G. (1989): The effects of temperature and host quality on the rate of increase of the grain aphid (Sitobion avenae) on wheat. *Annals of applied Biology* **115**: 3 – 9.

Anonymous (1952-1981): *Annual Meteorological Reports*. Meteorological Office, Bracknell, Berks.

Anonymous (1957-1980): *Annual Research Reports*. Department of Agriculture for Northern Ireland.

Anonymous (1984): *Winter wheat – managed disease control*. Leaflet 831 MAFF, ADAS Cereals Pathology Group.

Anonymous (1985): *Spring barley - managed disease control*. Leaflet 844 MAFF, ADAS Cereals Pathology Group.

Anonymous, (1989): *Animal Health and Production at Extremes of Weather*. WMO Technical Note 191, WMO No. 685, (Reports of the CagM Working Groups on Weather and animal disease and weather and animal health). World Meteorological Organization, Genève, Switzerland.

Anonymous, (1997): *Report of Massey University EpiCentre*, New Zealand, June 1997.

Audsley, E., Bailey, B.J., Beulah, S.A., Maddaford, P.J., Parsons, D.J. and White, R.P. (1997): Decision support systems for arable crops: increasing precision in determining inputs for crop production. *First European Conference on Precision Farming*, Warwick University, 8 – 10 September 1997.

Beaumont, A. and Staniland, L.N. (1933): *Ninth Annual Report of Seale Hayne Agricultural College*, Newtown Abbot, Devon.

Beaumount, A. (1947): The dependence on weather of the dates of potato blight epidemics. *Transactions of the British Mycological Society* **31**: 45 - 53.

Berkeley (1846): Observations, botanical and physiological, on the potato murrain. *Journal of the Horticultural Society of London* **1**: 9 – 34

Bouma, E. and Hansen, J.G. (1999): Overview of standard descriptions of Phytophthora Decision Support Systems In: H. Schepers and E. Bouma (Editors): *Proceedings of the workshop on the European Network for development of an integrated control strategy of potato blight*. Uppsala, Sweden, 4th – 13th September 1998. Applied Research for Arable Farming and Field Production of Vegetables, Lelystad, Netherlands.

Bourke, P.M.A. (1953): *The potato blight weather warning service in Ireland in 1952*. Technical Note 13. Irish Meteorological Service, Dublin.

Bourke, P.M.A. (1955): *The forecasting from weather data of potato blight and other plant diseases and pests*. WMO Technical Note 10. World Meteorological Organisation, Genève, Switzerland.

Bourke, P.MA. (1991): Potato blight in Europe in 1845: the scientific controversy. In J.A. Lucas, R.C. Shattock, D.S. Shaw and L.R. Cooke (Editors): *Phytophthora*. Cambridge Univerity Press, pp. 11 – 24.

Catney, D.C. (1997): *Mathematical Modelling of Abattoir Condemnation Data*. PhD Thesis, The Queen's University of Belfast.

Chaptal, J. (1923): *Les avertissments agricoles et la phytopathologie*. Congrès Pathologique végétale (Centinaire de Pasteur), Strasbourg.

Crosier, W. and Reddick, D. (1935): Some ecological relations of the potato and its fungous parasite Phytophthora infestans. *Phytopathology* **25**: 13.

de Bary, A. (1861): *Die gegenwärtig herrschende Kartoffelkrankheit, ihre Ursache und ihre Verhütung*. Felix: Leipzig.

de Bary, A. (1863): Rescherches sur le développement de quelques champignons parasites etc. *Annales des Sciences Naturelles* 4e sér. Bot. **20**: 1 – 148.

Donaldson, A. I., Lee M. and Gibson, C.F. (1987): Improvement of mathematical models for predicting the airborne spread of foot-and-mouth Disease. *Advances in Aerobiology* 351 - 354.

Duncan, A.J. (1998): Extensification in a cold climate: implications for thermal balance in sheep. In: Proceedings of a Workshop held in Aberdeen, Scotland, 5/6th March 1998, pp. 29 - 34.

EPPO (1978): *Guidelines for the biological evaluation of fungicides*. European and Mediterranean Plant Protection Organisation, Paris.

Everdingen, E. van (1926): Het verband tusschen de weergesteldheid en de aardappelziekte (P. infestans). *Tijdschrift Plantenziekten* **32**: 129 – 140.

Frost, M.C. (1994): *Phytophthora infestans on potatoes in Ireland - its incidence, control by fungicides and predictability in relation to weather*. Ph. D. thesis, National University of Ireland.

Fry, W.E., Apple, A.E. and Bruhn, J.A. (1983): Evaluation of potato late blight forecasts modified to incorporate host resistance and fungicide weathering. *Phytopathology* **73**: 1054 - 1059.

Gibson, T.E. (1978): The 'Mt' System for forecasting the prevalence of fascioliasis. In: *Weather and parasitic animal disease.* WMO Technical Note 159. WMO No. 497. World Meteorological Organization, Genève, Switzerland.

Gloster, J., Blackall R.M., Sellers, R.F. and Donaldson, A.I. (1981): Forecasting the airborne spread of foot-and-mouth disease. *The Veterinary Record* **110**: 47 - 52.

Goodall, E.A. and Menzies, F.D. (1995): Mathematical modelling for the control and prevention of animal production diseases. *Journal of Mathematics Applied in Business and Industry* **5**: 337.

Hadders, J. (1996): Experience with a late blight DSS (Plant-Plus) in starch potato area of the Netherlands in 1995 and 1996. *First workshop of a European Network for Development of an integrated control strategy of potato late blight.* Lelystad, The Netherlands, 30 September – 3 October 1996, pp. 117 - 122.

Hardwick, N.V. (1998): Disease forecasting. In D.G. Jones (Editor): *The Epidemiology of plant diseases.* Kluwer Academic Publishers, pp. 207-230.

Hope Cawdery, M.J., Gettinby, G. and Grainger J.N.R (1978): Mathematical models for predicting the prevalence of liver-fluke disease and its control from biological and meteorological data. In: *Weather and parasitic animal disease.* WMO Technical Note 159. World Meteorological Organization, Genève., Switzerland

Keane, T. (1986): *Climate, weather and Irish agriculture.* AGMET, Dublin.

King, J.E. (1972): *Septoria progress,* 1971. Open Conference of Advisory Plant Pathologists. Internal paper no. PP/0/207.

Large, E.C. (1959): The battle against blight. *Agriculture* **65**: 603-608.

Lindley, J. (1845): Editorial in *Gardeners' Chronicle* 23rd August.

Lowles, A.J., Tatchell, G.M., Harrington, R. and Clark, S.J. (1996): The effect of temperature and inoculation access period on the transmission of Rhopalosiphum padi (L.) and Sitobion avenae (F.) *Annals of applied Biology* **128**: 45 – 53.

Mackay, D.K., Lattuda, R., Verrier, P.J., Morris, R.S. and Donaldson, A.I. (1997): *EpiMAN (EU): Development of a computerised decision support system for control of outbreaks of foot-and-mouth disease within the European Union.* Final Report, CEC AIR Programme Contract No. AIR3 – CT92-0652.

MacCarthy, D.D., and Lynch, P.B. (1986): Animal Production and the environment. pp.182-213. In T Keane (Editor): *Climate Weather and Irish Agriculture.* AGMET, c/o Met Éireann, Dublin, pp. 329.

McIlroy, S.C., Goodall, E.A., McCracken, R.M. and Stewart D.A (1989): Rain and windchill as factors in the occurrence of pneumonia in sheep. *The Veterinary Record* 125: 79-82.

McIlroy, S.G., Goodall, E. A., Stewart, D.A., Taylor S.M. and McCracken, R.M. (1990): A computerised system for the accurate forecasting of the annual prevalence of fascioliasis. *Preventative Veterinary Medicine* **9**: 27 - 35.

Melhus, I.E. (1915): *Wisconsin Agriculture Experimental Station Research Bulletin.*

Mercer, P.C. and McGimpsey, H.C. (1985): Mildew of spring barley in Northern Ireland, 1982-84 II. Effect of timings and rates of fungicide sprays. *Record of Agricultural Research (Department of Agriculture, Northern Ireland)* **33**: 43 – 48.

Mercer, P.C. and Ruddock, A. (1999): The interaction of variety and fungicide dose in the control of mildew of spring barley. *Proceedings of Crop Protection in Northern Britain*, Dundee March 1999, 61 – 66.

Mercer, P.C., McGimpsey, H.C. and Malone, J.P. (1986): Surveys of diseases of spring barley in Northern Ireland. *Record of Agricultural Research (Department of Agriculture, Northern Ireland)* **34**: 17 – 27.

Millardet, P.M.A. (1885): Traitement du Mildiou par le Melange de Sulfate de Cuivre et de Chaux, etc. *Journal d'agricultural pratique* 1885 (2). 513 – 516, 707 - 710, 801 – 805.

Morren, C. (1845): Notice sur le Botrytis devastateur ou le champignon des pommes de terre. *Annales de la Société Royale d'Agriculture et de Botanique de Gand* 1: 287 – 292.

Mount, L.E. and Brown, D. (1982): The use of meteorological records in estimating the effects of weather on sensible heat loss from sheep. *Agricultural Meteorology* **27**: 241 - 255.

Mount, L.E. and Brown, D. (1983): Wind chill in sheep: its estimation from meteorological records. *Agricultural Meteorology* **29**: 259 - 268.

O'Brien, D.J. (1989): *Cold weather and animal disease. In: Animal health and production at extremes of weather.* WMO Technical Note 191. World Meteorological Organization, Genève, Switzerland, pp. 45 –5 0.

Ollerenshaw, C.B. (1966): The approach to forecasting the incidence of fascioliasis over England and Wales 1958-1962. *Agricultural Meteorology* 3: 35 - 53.

Ollerenshaw, C.B. and Rowlands, W.T. (1959): A method of forecasting the incidence of fascioliasis in Anglesey. *The Veterinary Record* **71**: 591 - 598.

Olsen, O.W. (1947): Longevity of metacercariae of Fasciola hepatica on pastures in the upper coastal region of Texas and its relationship to liver fluke control. *Journal of Parasitology* **31**: 36 - 42.

O'Neil, T.P. (1946): The scientific investigation of the failure of the potato crop in Ireland 1845 – 6. *Irish Historical Studies* **5**: 123 – 138.

Parker, S.R., Royle, D.J. and Hims, M.J. (1994): *Control of Septoria and eyespot in winter wheat: fungicide applicxations in response to growth stage and forecasts of disease risk.* Project Report 95 Home Grown Cereals Authority, Hamlyn House, Highgate Hill, London.

Pasquill, F. (1974): *Atmospheric Diffusion – The Dispersion of windborne material from industrial and other sources* (2nd edition). Ellis Horwood Limited, Chichester (distributed by John Wiley & Sons, London).

Paveley, N.D. (1998): Integration of epidemiology, crop protection and physiology as a biological basis for decision support. *Proceedings of International Congress of Plant Pathology*, Edinburgh August 1998, paper 3.1.1S

Paveley, N.D., Lockley, K.D., Sylevester-Bradley, R. and Thomas, J. (1997): Determinants of fungicide spray decisions for wheat. *Pesticide Science* **49**: 379 – 388.

Polley, R.W. and King, J.E. (1973): A preliminary proposal for the detection of barley mildew infection periods. *Plant Pathology* **22**: 11 – 16.

Polley, R.W. and Smith, L.P. (1973): Barley mildew forecasting. *Proceedings 7th British Insecticide Fungicide Conference* , 373 – 378.

Sanson, R.L. (1994): EpiMAN – A decision support system for managing a foot and mouth disease epidemic. *Surveillance* **21**: 22 – 24.

Secher, B. and Bouma, E. (1996): *Survey on European Crop Protection Decision Support Systems*. SP report no. 16 Vol. 4. Danish Institute of Plant and Soil Science.

Smith, L. P. (1956): Potato blight forecasting by 90 per cent humidity criteria. *Plant Pathology* 5: 83 - 87.

Sørensen, J.H. and Christian Ø.J. (1996): *A computer system for the management of epidemiological data and prediction of risk and economic consequences during outbreaks of Foot-and-Mouth disease*. CEC AIR Programme Contract No. AIR3 – CT92-0652 (Danish contribution to Final Report). Danish Meteorological Institute, Scientific Report 96 - 7, Copenhagen.

Sparks, W.R. (1980): *A model relating the progress of potato blight to weather*. Agricultural Memorandum 899, British Meteorological Office, Bracknell, UK.

Sparrow, L.A.D. (1974): Observations on aphid populations on spring-sown cereals and their epidemiology in south-east Scotland. *Annals of applied Biology* 77: 79 – 84.

Starr, J.R. (1981): Weather and lamb mortality in a commercial lowland sheep flock. *Agricultural Meteorology*, 24: 237 - 252.

Starr, J.R. (1984): *Operational forecasting of a 'wind-chill' factor for young lambs*. The Meteorological Magazine 113 (1342), May 1984, Meteorological Office, Bracknell, UK, 105 - 113.

Starr, J.R. (1988): *Weather, climate and animal performance*. WMO Technical Note 190, WMO No. 684, World Meteorological Organization, Genève, Switzerland.

Thomas, R.J. (1978): Forecasting the onset of nematodiriasis in Sheep. In: *Weather and parasitic animal disease*. WMO Technical Note 159. World Meteorological Organization, Genève, Switzerland.

Thomas, M.R., Cook, R.J. and King, J.E. (1989): Factors affecting development of Septoria tritici in winter wheat and its effect on yield. *Plant Pathology* 38: 246 - 257.

Tyldesley, J.B. and Thompson, N. (1980): Forecasting Septoria nodorum on winter wheat in England and Wales. *Plant Pathology* 29: 9 – 20.

Ullrich, J. and Schrödter, H. (1966): Das Problem der Vorhersage des Auftretens der Kartoffelkrautfäule (Phytophthora infestans) und die Möglichkeit seiner Lösung durch eine "Negativprognose". *Nachrichtenblatt Deutsches Pflanzenschutzdienst* (Braunschweig) 18: 33 - 40.

Vickerman, G.P. (1977): Monitoring and forecasting insect pests on cereals. *Proceedings 1977 Brighton Crop Protection Conference – Pests and Diseases* (Vol. 1) 227 – 234.

Wall, R., French, N.P. and Morgan, K.L. (1993): Predicting the abundance of the blowfly Lucilia sericata (Diptera: Calliphorida). *Bulletin of Entomological Research* 83: 431 - 436.

Watson, S.J. and Carter, N. (1983): Weather and modelling cereal aphid populations in Norfolk (UK) *EPPO Bulletin* 13: 223 – 227.

Zadoks, J.C., Chang, T.T. and Konzak, C.F. (1974): A decimal code for the growth stages of cereals. *Weed Research* 14: 415 – 421.

Chapter 8   Weather Forecasting by Numerical Weather Prediction

Tom Sheridan
Met Éireann, Glasnevin, Dublin, Ireland

8.1   Numerical Weather Prediction

Farmers have always been aware of the impact of weather on their lives. The agricultural sector is acutely conscious of the need for, and the potential benefits of, accurate weather prediction. The folklore of almost every society is replete with rules-of-thumb regarding weather intended to aid farmers in the all-important decisions they must make on an almost daily basis ("Birds flying low, expect rain and blow"). In the early 20th century, with improving observations, conceptual models of the atmosphere (introducing the now-familiar concept of fronts) were developed. It was realised that Newton's laws and the laws of thermodynamics could form a complete set of predictive equations, and weather prediction moved into its scientific phase. Meteorological scientists knew that the implication of this realisation was that weather forecasting could be treated as an initial value problem. It was recognised that a tremendous amount of calculation was involved, but the first attempt to predict the weather numerically (involving years of work with mechanical calculators to produce a six-hour forecast!) produced results which were completely unrealistic. No further attempts at Numerical Weather Prediction (NWP) were made for almost three decades, until the 1950s, because few tools were available to further develop or use the mathematical models. We had to depend on the skill and knowledge of meteorologists to forecast our weather. These scientists produced (and continue to produce) forecasts with considerable skill for a time horizon up to about two days ahead. Table 8.1 shows some of the highlights in the development of modern meteorology and their effects. Some of the earliest applications developed for computers were in the area of Numerical Weather Prediction (NWP) and by the early 1960s computer models were producing output which soon became essential to forecasters, particularly in their attempts to improve predictive skill beyond 36 hours. Since that time, much effort has been devoted by meteorologists to improving NWP output.

The aim of this chapter is to explain NWP computer models of the atmosphere in a way which can be understood by the non-meteorologist, and to provide an indication of the strengths and weaknesses as input data for the agricultural decision making processes. For readers who would wish to study the topic presented in greater depth there are several textbooks available. Good examples include Stull (1995) – a good general description of meteorology for the scientifically minded non-meteorologist (Chapters 9 and 14 of particular relevance to the current topic) and Haltiner & Williams (1980) – a standard textbook for meteorologists working in the NWP area.

Following a general description of NWP, a little detail is provided on the input data required, the initial analysis stage, the prediction stage and model types. Subsequent sections deal with the question of model resolution, output parameters, accuracy/errors and availability/format of NWP output. In the final section, some indication is given of future developments in the area.

Table 8.1: Important developments in meteorology over the last 150 years

Date	Developments	range of predictions
Ca. 1850s	Development of telegraph; establishment of national meteorological services	12 hours
1900 - 1920	Theory of fronts	24 hours
1930s and 40s	Expansion of Upper air observations	36 hours
1960s	NWP; increases in computer power	2 days
1970s	Improved NWP models	3 days
1980s	NWP; satellite derived products	4 days
1990s	NWP; ensemble methods; improved data assimilation	5 days
2000- 2010	NWP; future technology…	7 days

8.1.1   NWP - What is it?

While we commonly think of the atmosphere as flowing horizontally over the surface of the earth, in reality the air is also continually rising and falling. This vertical motion is most important from the point of view of evaporation and condensation of the moisture in the air and for the generation of various types of precipitation (e.g. rain, sleet, hail, snow). Thus any attempt at weather prediction must take a three dimensional view of the atmosphere. As the atmosphere is a compressible fluid, its motion can be described by the hydrodynamic equations for fluid flow.

In their simplest form, these equations use six variables: wind speed (eastwards, northwards and vertical), pressure, temperature and density (or specific volume). Of these, the horizontal wind components, pressure and temperature can be directly observed while the remaining two variables are inferred from the governing equations.

A simple mathematical model of the atmosphere would have six equations describing the change of these variables with time.

8.1.2   The Governing Equations

The first set of equations is based on Newton's second law of motion (Force = mass times acceleration) and relates the rate of change of the velocity (in the three dimensions) to the forces acting on the air mass. In vector form these three equations can be stated as:

$$\frac{d\vec{V}}{dt} = -\alpha\vec{\nabla}p - 2\vec{\Omega}x\vec{V} + \vec{g} + \vec{F}$$

$$8.1$$

where

$\vec{V}$ is the velocity of the air (in three dimensions: u, v, w in the eastwards, northwards and vertical directions)

α is the specific volume (the inverse of density)

p is the pressure

$\vec{\Omega}$ is the angular velocity of the earth

$\vec{g}$ is the sum of gravitation and the centrifugal force

$\vec{F}$ is the friction force.

The first term on the right represents the force due to the gradient of pressure, the second term (known as the Coriolis force) is due to the rotation of the earth, the third term is the sum of the gravitational and centrifugal force and the fourth term represents the frictional force. The first term causes air to accelerate towards low pressure. The second term causes air to accelerate to the right (in the Northern Hemisphere), until an approximate balance of the Coriolis and pressure gradient forces occurs; a consequence of this is that, in the balanced state characteristic of large scale weather systems, the air circulates anti-clockwise around low pressure systems and clockwise around high pressure systems. The third term causes the air to be drawn towards the centre of the earth (modified slightly by the tendency to move away from the earth's axis due to rotation of the earth). The frictional force tends to dissipate the energy of the moving air.

The next equation represents the conservation of mass and is used to calculate the rate of change of air density:

$$\frac{1}{\alpha}\frac{d\alpha}{dt} = \vec{\nabla} \bullet \vec{V} \qquad\qquad 8.2$$

where the term on the right is $\dfrac{\partial u}{\partial x} + \dfrac{\partial v}{\partial y} + \dfrac{\partial w}{\partial z}$, the divergence of velocity.

The relationship between pressure, density and temperature is represented by:

$$p\alpha = RT \qquad\qquad 8.3$$

where R is the gas constant for air, T is the temperature (Kelvin). This so-called equation of state relates the thermodynamic variables for a gas which obeys Boyle's and Charles' laws.

Finally, the effect of addition or subtraction of heat is represented by:

$$Q = c_p \frac{dT}{dt} - \alpha \frac{dp}{dt} \qquad\qquad 8.4$$

Where Q is the rate of heat addition, c_p is the specific heat at constant pressure. This equation represents the first law of thermodynamics and expresses the principle of conservation of energy.

The above six equations (Equation 5.1 being equivalent to three scalar equations) can be used together to calculate the rate of change with time of the dependent variables (u, v, w, p, α, T) as a function of their rate of change in space. Thus they can be used to predict the values of the variables at any particular time. Of course the real world is more complex than this – equations for moisture must be included and many other physical processes must be taken account of (such as cloud formation and radiation effects either explicitly or by parameterisation), but the basis from which all NWP is developed is this set of equations.

However, due to their complexity, these second order partial differential equations cannot be solved analytically but must be integrated by approximate methods. In order to convert the differential system into a simpler algebraic system, the atmosphere is divided into small boxes and the dependent variables within each box are represented by their values at the centre of the box. Thus, for example, dy/dx is represented by the difference Δy between the y values in adjacent boxes divided by the distance Δx between box centres. The equations must be solved at each point on a fairly closely spaced grid over the earth and in the vertical. Any computational model must take account of the conditions at the lateral boundaries of the area in question, its upper boundary and at the surface of the Earth. For computational reasons the solution will produce predicted values of the variables only at a very short time step (typically only 10 or 15 minutes) into the future. These values serve as a starting point for a further calculation to bring the predictions to the next time step. Thus there is a huge amount of calculation to be done, even to produce a 24-hour forecast – hence the need for high-speed computers. Such a simple model can produce values for the six parameters at the designated grid points which are then interpreted by the forecaster. While the automatic output could be routed to customers, a skilled meteorologist examining all of the output and other information can add value to the products both in terms of improving quality and adding other variable not directly output by the NWP.

Modern NWP systems are considerably more complex because they take account of the interactions of the air with the earth's surface, the effects of cloud and so on. Even if we are interested in a relatively limited area (such as Ireland), the model will need to do the calculations over a very extended area. This is because developments taking place thousands of kilometres away can quite often effect our weather within a couple of days. Figure 8.1 shows average eastward speed of movement of the air on the north Atlantic at a level of about 5.5km (contour labels in units of metres per second – double the values to convert to knots) during January (ECMWF, 1997a). Developing weather systems would normally move considerably faster than this and hence the need for the NWP domain to cover a relatively large part of the globe.

8.1.3   Input Data

The most important input observation for NWP is pressure, or equivalently, the height of constant pressure surfaces. Wind, temperature and humidity are also required. Many other meteorological variables are routinely being observed and some are now being incorporated into the models. The primary sources for the data are the synoptic stations (operated by each National Meteorological Service –NMS), ship reports and radiosonde reports. The latter are from hydrogen-filled balloons released several times per day from stations around the earth with sensing capability for wind (speed and direction), pressure, temperature and humidity. They provide profiles of these variables through the atmosphere to distances of several tens of kilometres above the surface (over 500 reports at 1200 Co-ordinated Universal Time (UTC) on a typical day).

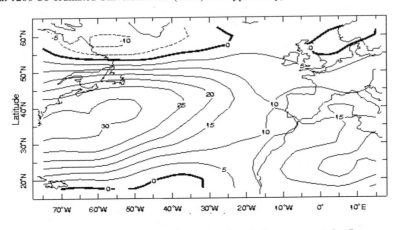

Figure 8.1: Typical 500 hPa mean westerly wind component in January (ECMWF)

One of the biggest problems in weather prediction is the sparseness of weather observations. Even with comprehensive meteorological observation stations blanketing the earth at intervals of a few kilometres, forecasters will be quick to point out that the variations in between stations will sooner or later cause even the best forecast to degrade. NWP experts would be very happy to have input data spaced at 10-30 km intervals. The reality, however, is that the spacing between observing stations is on average hundreds of kilometres and (of critical importance to Ireland) much greater over the oceans. Of the 12,000 or more surface observations at 1200 UTC on a typical day, less than 10% are from ships (Figure 8.2). The situation has improved in recent years with the availability of more weather reports from ocean buoys (1,400 at 1200 UTC on 30/12/98); atmospheric temperature soundings from orbiting satellites (over 2,100 reports on the sample date and time, albeit of much coarser vertical resolution than available from the 500 radiosoundings) and cloud motion winds and upper tropospheric humidity reports from geostationary satellites (95,000). Recent developments in satellites are improving the situation regarding sea level reports however, with wave, pressure, wind and temperature data as well as water vapour reports through the depth of the atmosphere now becoming available from these remote sensors.

8.1.4   Analysis

Figure 8.3 shows schematically how NWP fits into the suite of products provided by National Meteorological Services to the agricultural community. Meteorological reports from around the world are rapidly disseminated to the NMSs where they are decoded and quality controlled. As well as being plotted and analysed by forecasters, they are used to determine the appropriate initial values at the regularly spaced grid points required by NWP. A sophisticated scheme is then used to blend the observations with a "first guess" analysis derived from a previous short-range forecast to produce the best initial field for the automatic prediction scheme. The analysis incorporates as much as possible meteorological data (including reports that do not coincide precisely with the nominal analysis time). Appropriate weights must be assigned to different observation types and erroneous reports must be eliminated or the errors must be corrected. The initialisation process ensures that the atmosphere is in balance at the start of the prediction process in order to avoid the development of spurious waves. Particular attention must be paid to the boundaries of the geographical area under consideration as errors at the boundaries can quickly contaminate the whole domain.

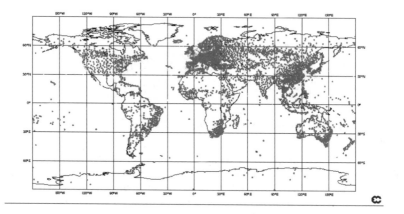

Figure 8.2: Data coverage - observations for 1200 on 30/12/98 (ECMWF)

8.1.5   Prediction

Once initialised, the NWP model calculates values for each variable at each grid point, step by step, as far ahead as is required. Even with a fairly limited model, the amount of computation is huge. For example, a limited area model with 200 grid points in the east-west direction, 150 points in the north-south direction, 20 levels in the vertical and a 15 minute time step will need all the calculations to be done at each of about 120,000,000 points in space and time to produce a two day forecast. Older models used values (often derived from climatology) for sea surface temperature, surface roughness, snow cover and the earth's surface moisture, which were held constant during the model run. These parameters change dynamically in the course of a run of most of today's models. Modern NWP also treats the interaction of clouds and radiation and the drag of orography on the atmosphere in a much more realistic way. For example, many modern models treat explicitly the fact that mountains can cause waves in the atmosphere (sometimes visible as lenticular clouds) leading to momentum exchange between the earth and the atmosphere.

8.1.6   Post processing

At the end of each model run there will be a forecast of the basic variables along with other forecast information such as rainfall and cloudiness, some of which will be explicitly part of the predictive scheme and others which will be derived for the relevant points. These direct model outputs can be made available to customers outside the NMSs as indicated in Figure 8.3. Often they will be further processed by statistical methods, such as Kalman filters or model output statistics methods, to improve the forecast values (e.g. to remove known bias in the values) and these are displayed on workstations for further modification by forecasters who produce the final products for dissemination to the customer.

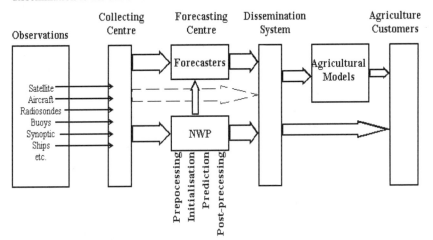

Figure 8.3: Data flow from observation to agricultural community

8.1.7   Model Types

Global

The impact of distant atmospheric activity means that, if one wishes to produce forecasts for any particular location beyond two or three days ahead, it is essential to run a model that covers the whole planet. Another advantage of running a global NWP model is that there are no horizontal boundaries which makes the model formulation somewhat easier. There is, of course, a cost: the amount of computation is huge. For this reason, European NMSs have set up the European Centre for Medium-range Weather Forecasts (ECMWF, 1997a). This Centre routinely upgrades its computers to ensure it continues to have leading edge technology. Its current atmospheric model runs on a supercomputer with 116 parallel processors together achieving 250,000,000,000 operations per second. Global models tend to have coarser resolution than their limited area counterparts but have better ocean-atmosphere interaction, which is of increasing importance as we attempt to predict weather patterns further into the future. Since mid-1998 ECMWF runs a coupled atmospheric and ocean wave model and initial indications are that this has added some hours to the useful range of prediction. Figure 8.4 shows a typical extended range forecast product prepared in Ireland from the ECMWF global model.

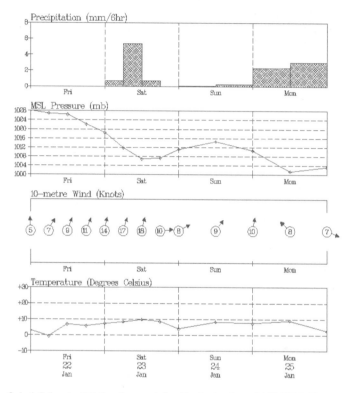

Figure 8.4: 1.5 day to 6.5 day forecast for Cork city from ECMWF Global model

Limited Area

In addition to accessing a global model, most NMSs run NWP on a limited area, normally centred on their own country. This permits them to use a higher grid resolution and to optimise the coding and output to their own requirements. Also, other application models (e.g. models to produce wave predictions or pollution dispersion models) can be integrated with or can be run using the output of the local NWP model. Some of these models have a variable grid spacing with highest resolution centred on one's own country. Rather than having the weather variables fixed at the boundaries, most limited area models use predictions from global models to determine the values at the boundaries. A typical modern limited area model as described by Puri *et al.* (1998) uses such methods. Smaller NMSs tend to develop their models in co-operation with each other, thus bringing cross-fertilisation of ideas and avoiding duplication of research effort. The High Resolution Limited Area Model (HIRLAM) was developed by co-operation between the NMSs of 8 countries including Ireland and is the only atmospheric NWP model in use in Ireland. The system is described in HIRLAM newsletters published several times per year (copies available from Met Éireann). In the case of HIRLAM, each member country implements its own version of the model and some (e.g. Denmark) run it in nested form with very high resolution over small areas of particular interest. Often such models are used for predictions up to 48 hours ahead of data time, occasionally as far as 72 hours ahead.

Mesoscale
Mesoscale models are run over very limited areas on grid spacing of 10km or less. They are normally much more sophisticated in terms of formulation than the larger scale models. They are particularly good at predicting wind flow in rough terrain, sea breezes and high-resolution short-range rainfall. At the time of writing such models are mainly in the testing phase in the academic domain. Mesoscale models with resolutions of 2km have been used operationally for exceptionally weather sensitive events (Snook *et al.*, 1998). As the number of automatic weather stations increases and remote sensing technology improves, the higher resolution input data to drive such models will become available to NMSs. Increasing computer power and falling prices are causing the distinction between mesoscale and limited area models to disappear.

Seasonal
Long-range prediction models (ECMWF 1997b) differ considerably from those used for predictions of the order of a week ahead. The oceans have a major influence on the atmosphere (and vice versa) and changes in ocean currents, temperature and salinity induce changes in the atmosphere. Such changes take place relatively slowly and so they can be safely ignored in short range NWP, but they become increasingly important when atmospheric models are run for the longer range. It is also recognised that deterministic forecasting (in the sense of a description of the day to day variations at a given location) is not possible beyond a time horizon of about two weeks. Therefore, ensemble methods are the norm in this case.

The output of seasonal prediction systems is normally in a different format from that of other NWP. Typically they provide output in probabilistic form for periods of months grouped together. Output is often expressed in terms of its relationship to climatologically normal figures. For example, they may provide probabilities of rainfall or temperature being above or below normal for the season in question. Also, there are a considerable number of occasions, particularly in extra-tropical regions, where the results are too divergent to provide any guidance (white areas in Figure 8.5).

Climate models used in the investigation of climate change scenarios differ even more from the other models described here – for example, they include more sophisticated treatment of chemical processes and often employ assumptions regarding future concentrations of greenhouse gases – and are beyond the scope of this chapter. For a recent review of such models see Carson (1999). However, the outputs of global climate models is sometimes used as input to the 'normal' NWP-type models (such as HIRLAM) in order to see more clearly the regional or local effects of possible climate change.

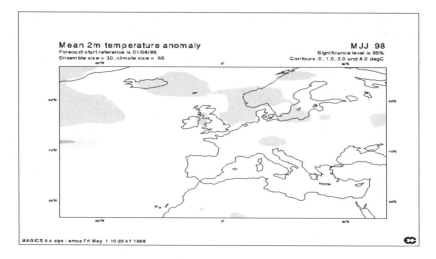

Figure 8.5: Three month prediction of temperature departure from normal. Darker shading near Ireland and the N. Sea indicates temperatures are expected to be 1 to 2 ^{0}C above normal

8.2   NWP Resolution

8.2.1   Temporal

The time range to be covered by the forecast is the main determinant of the frequency of model runs. Typically seasonal models are run once or twice per month, global models once per day and limited area models four or more times per day. Theoretically, all models could output results at every single time-step, but (ignoring seasonal models) output is normally confined to 6 or 12-hour intervals with rainfall accumulated between these times also being available. NMSs running their own models (e.g. HIRLAM, Figure 8.6) have the flexibility to produce output for a limited number of grid points, at much finer temporal resolution and may produce outputs such as daily maximum and minimum temperature forecasts, which depend on monitoring all time steps.

8.2.2   Horizontal

Horizontal resolution is steadily improving. During the past couple of decades, resolution has improved typically by a factor of 8 to 10. Currently 20 to 40 km spacing is the norm, with a number of the limited area models now running below 20 km. Limited area models now have upwards of a hundred grid points over an area the size of Ireland (Figure 8.7). The latest mesoscale models have resolutions considerably below 10 km, but for optimum benefit require higher resolution input data than is generally available. Note: most models have 30 to 50 levels in the vertical but, while higher resolution in the vertical contributes substantially to the overall output quality, vertical resolution (apart from that in the boundary layer, close to the ground) is of little direct interest to agricultural users.

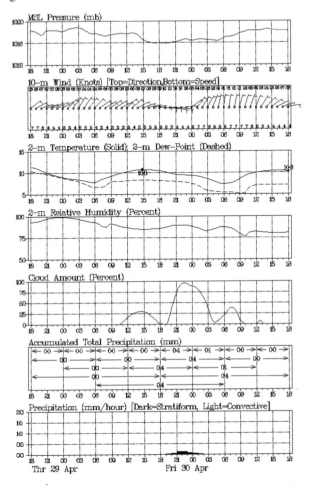

Figure 8.6: Met Éireann HIRLAM forecast for Dublin showing improved temporal resolution compared to Figure 8.4

8.2.3   Topographical

The weather at a particular location is dependent on height above sea level and on the local topography. This can be clearly seen in, for example, rainfall patterns where accumulations can be considerably higher on windward slopes than in the lee of relatively small mountains. The surface relief used by the NWP has to be smoothed considerably to avoid spurious results. This gives rise to a serious lack of information on

local variations, which is of particular interest to the agricultural sector. The effects of better horizontal resolution on forecast rainfall accumulations can be seen in Figure 8.8, produced in the course of experimentation within Met Éireann. As the horizontal resolutions of models improve, more realistic topography has been introduced and this trend is continuing but to date, for some locations post processing of output, preferably based on empirical observation at a location will be required where topography varies considerably.

Figure 8.7: 30-km resolution grid-points (IRL HIRLAM 1998) showing the grid resolution relative to the size of Ireland

8.3 NWP Output Parameters

The output variables from a typical modern NWP model (at the appropriate time steps) are:

- Mean Sea Level Pressure (hPa, of direct use for certain pressure sensitive instrumentation but generally used to determine the general weather patterns)
- Wind Speed at 10 m above the surface (m s^{-1} or knots, important for spraying decisions, evaporation calculations)
- Wind Direction (usually degrees from north, from which the wind is blowing)
- Air temperature at 2 m above the surface (^{0}C, impacts growth rates and drying conditions)
- Dew Point at 2 m above the surface (^{0}C, with temperature indicates humidity of the air)
- Total precipitation (often mm hr^{-1} or accumulation in mm since the previous output step of the model)
- Total cloud cover (percent or eighths of sky covered)
- Radiation (watts m^{-2}, solar and thermal, at the earth's surface)
- Evaporation from the earth's surface (mm hr^{-1})

- Relative humidity at 2 m (percent, can be derived from temperature and dew point)
- Separate values for shower type precipitation and precipitation of a more persistent nature
- Low cloud cover (percent or eighths of sky covered)
- Conditional probability of snow or hail (percentage probability, assuming some precipitation is falling)

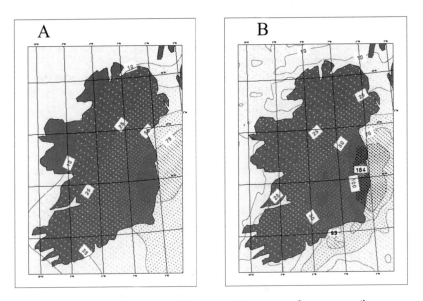

Figure 8.8: A. Rainfall Prediction (mm) from 0900 25th to 0900 26th August 1986 with 55km grid. A. Vs B, with 10 km grid. A maximum of 190mm was recorded in Wicklow

8.4   NWP Accuracy

In general terms, since the 1950s, there have been steady improvements in the accuracy of NWP output. It should be emphasised that in what follows comments refer to direct NWP output. While model output continues to improve, it is a fact (and is likely to continue so for some time to come) that statistical post-processing and/or forecaster intervention still adds considerably to the skill and value of forecasts. Great strides have been made in improving accuracy and skill, but as accuracy for a given lead time improves it becomes more difficult to make further advances. Practical and theoretical studies indicate that marginal quality increases in high quality forecasts bring disproportionately large improvements in the *value* of the forecast (see, for example, Katz, 1987). For further information on the question of the value of weather information, see Wilks, 1997.

Various measures of accuracy and skill (i.e. accuracy compared with some naïve forecasting method such as persistence or climatology) are used in the world of meteorology. Amongst the more common measures are:

- Bias or mean error. Modern NWP output has very small errors for most variables when aggregated over a large area. However, significant bias can still exist for particular locations or in certain conditions (e.g. temperature biases where surface vegetation or snow cover differs from that in the model)
- Mean Absolute Error (MAE) which is the average error where the sign (plus or minus) of the error is ignored
- Root Mean Square Error (RMSE) which is the square root of the average of the squared errors. This is related to variance and is one of the most useful measures for further statistical manipulation.
- Anomaly Correlation Coefficient (ACC) which is a measure of the correlation between the deviations from climatology of the forecast and observations, and thus is a measure of model skill by comparison with climatology.

8.4.1   Types of errors/shortcomings

Inaccuracies in NWP result from initial observation or analysis errors (e.g. from lack of data, undetected data errors), smoothing in the analysis stage, the approximate nature of the orography and the approximations used to solve the equations. Forecast errors result, for example, from failure to detect deepening lows due to an incorrect initial analysis. Sometimes the sequence of development can be correctly predicted, but the timing is in error which, if the timing is greatly incorrect, is as unwanted as an incorrect prediction of developments.

8.4.2   Lead time and 'useful' lead time

A convenient definition of useful lead-time is the time (starting from the analysis time, t=0) for accuracy or skill to degrade to a level approaching that resulting from the use of climatological normal values. At that lead time a user might just as well use the (30-year) mean value for the variables. A value of ACC in excess of 60% is often taken as the delimiter of a 'useful' model forecast. To a degree, the range to which forecasts remain useful depends on the application. To illustrate this concept, two examples are:

- If an indication of above normal rainfall averaged over a three-day period is required there will often be useful information in the NWP not only during the first five days but for days six, seven and eight. Predictions for the individual days or, say, 12-hour periods at lead time of five or more days come with a low accuracy expectation. Such considerations have obvious implications for harvesting and haymaking operations, for example.
- Wind speed and direction at 500 hPa (about 5.5 km above ground level) for aviation planning will be accurately predicted to a longer lead time than is possible for wind at 2 m for agriculture or at 20 m for the construction industry.

The useful lead-time will also be dependent on the meteorological parameter in question.

8.4.3 Growth of errors with lead time

Chaos theory (Lorenz, 1963) indicates that there is a limit to the deterministic predictability of the atmosphere. Even with perfect model formulations and well-distributed and accurate reports the theoretical limit of predictability is estimated to be approximately two weeks (Lorenz, 1982). However, with the present state of both modelling and observations, skill levels of deterministic models drops below useful levels well before that time. Certain initial states of the atmosphere are more 'unstable' than others in that even very minor differences in starting conditions will lead to very substantially different final states at an even earlier stage.

It is important to note that timing errors alone increasingly degrade forecasts as time goes by. For example, if there is a 10% error in time of arrival (or, equivalently, in the speed of movement of weather systems), a forecast of 3-hour rainfall accumulation can be seriously in error by 1 ½ days, though 12 hour accumulations may retain useful skill to near 5 days. As well as error due to the speed of movement of weather systems, as we go further into the forecast period we have an increasing possibility of errors due to developments which have their origins below the scale of the original analysis grid. Most of the effort of the NWP community over the past thirty years has been focused on the problem of extending the range to which useful forecasts can be produced. As an approximation, the useful range has been extended by about one day per eight years over most of the period – i. e. forecasts for six days ahead in 1992 were about as good as those for five days in about 1984, or four days in the late 1970s. During the early 1990s the world's leading NWP models showed only small improvements in the verification scores, though in more recent times scores are again improving.

NWP produces useful information over a time range which varies geographically. Seasonal mean forecasts for the Tropics have some skill out to a range measured in weeks or even months. In the free atmosphere (i.e. well above the direct influence of the earth's surface) in Europe, NWP output is useful for a time period of between six and eight days (Figure 8.9). Close to the surface of the earth, however, for most weather parameters NWP skill drops below acceptable levels somewhat sooner.

8.4.4 Magnitude of errors for different parameters:

In most cases the bias (i.e. average error) for information such as wind speed, temperature and cloud is very small and remains low throughout the forecast period of the particular NWP model. However, the spread of forecast values above and below the verified values increases with longer lead times. Many of the values quoted in the following are from ECMWF, 1997c.

Temperature at 2 or 1.5 metres
Limited area models typically have RMSEs of about 2 °C initially, disimproving to between 2.5 and 3.5 °C at 48 hours. The ECMWF global model is widely used in Europe, though it suffers from the disadvantage that it runs only once per day and its output is available later than the limited area models run from the same data time. Most European countries perform verification of its output (mainly over their own territories). In Germany the percentage of cases with temperature within 2 °C of observed is about 80% in the early stages, 60% at 3 days. RMSEs have been found to be about 1.6 °C on

day 1 rising to 3.6 on day 7 (Iceland), 2.6 on day 1 rising to 4.2 on day 10 (Portugal), 1.9 day 1 reaching 3.4 °C by day 10 (Spain). In Denmark, the MAE has been found to increase from about 1.3 °C on the first day to about 2.4 °C on day 6. For Ireland, at an inland location, the bias of the 60 hour ECMWF forecasts (verifyed at midnight) is -0.95 °C (i.e. forecasts a little too cold) and MAE is 1.8 while the 72 hour forecasts (verifyed at 1200 UTC) are on average too warm by 0.44 °C with an MAE of 1.5 °C. Much of the variability of accuracy between different countries is due to the difference in variance of the observations between different geographic locations.

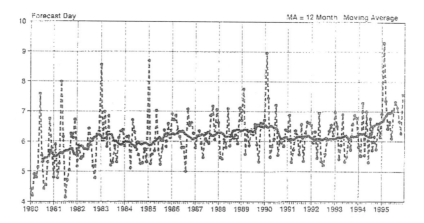

Figure 8.9: Forecast range at which the monthly mean 500 hPa anomaly reaches the 60% value over Europe for ECMWF predictions for each month (Dashed) and 12-month running mean (solid). From Simmons (1996)

Wind at 10 metres
Forecast mean wind at 10 m is available from most NWP models. Errors vary widely depending on precise location. Local topography and surface roughness have a major bearing on the local wind, whereas representation of these in models is by single values each representing an area the size of the square of the grid-point distance. Within Europe reports from a variety of countries indicate forecast wind speed within 2 knots (about 1 m/sec) of observed speed in about 60% of cases at the early forecast stages, 50% at 3 days. In Finland the 10 m RMSE vector error was found to be 4 knots (about 2 m/sec) at 6 hour becoming 4.5 knots at 48 hr. The RMSE vector wind error is found to be 7 knots at 6 hr, becoming 9 knots at 48hr in Norway, The Netherlands and Ireland. RMSE for wind speed is about 5 knots in Spain, Norway and Ireland (6 to 48 hour forecasts) [*HIRLAM Newsletter* **31**]. In the longer range, 24-hour mean wind speed in Denmark is found to have a MAE varying from 2 knots on day 1 to about 4 knots on day 6.

Rainfall
Improvements in NWP rainfall predictions have lagged behind that of other variables. This is explained, in part at least, by the fact that many of the processes that affect precipitation developments occur on scales smaller than model grid resolution. In general, NWP is found to over predict small amounts of rainfall. Short period

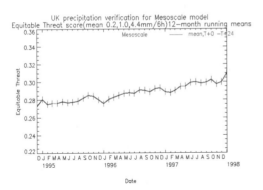

Figure 8.10: Time series of monthly equitable
threat scores for precipitation using the
mesoscale model at various forecast ranges.
The score is an average of the three separate
scores using thresholds of 0.2, 1.0 and 4.0 mm
hr⁻¹. The verification is against representative
UK observing stations

accumulations   of
precipitation tend to be
poorly forecast. Against this,
accumulations over 24 hours
tend to have considerable
skill.   For   example,
verifications of ECMWF
forecasts in Ireland indicate
rainfall category (three
categories: < 0.3, 0.3 to 5.0,
> 5.0 mm per 24 hours) to
be correctly predicted about
60% of the time out to 72
hours or more ahead. Figure
8.10 shows improvements
attained over 3 years in
precipitation forecasting in
one NWP centre. Note: the
Threat score refers to the
number of correct forecasts
of the category in question
divided by the sum of all
forecasts of occurrence and
occurrences not forecast.

Other Parameters
Verification studies are available for a range of parameters where the forecasts have
been produced by meteorologists. Generally speaking, however, less work has been
done in the European context in verifying NWP forecast surface parameters other than
those described above. Isolated studies indicate current cloudiness NWP forecasts within
25% of observed in about 70% of cases in early forecast stages, 60% at 3 days; 2 m
relative humidity (expressed in percent) RMSE generally within 13 percentage points of
observed in the 6 to 48 hours range.

8.5   NWP data availability/format

The range of meteorological variables output directly from the NWP models (section 8.3
above) and derived variables, combined with improved accuracy (section 8.4), means
predictive agricultural models are becoming ever more useful operationally. Growth
models for sugar beet or winter wheat, for example, can now use NWP predictions of
solar radiation as well as temperature, humidity and rainfall. Many agricultural
predictive models such as grass growth models which traditionally used current and past
meteorological data can now incorporate the increasingly reliable NWP forecasts of
sunshine, rain and temperature for several days ahead. Grass and silage quality, soil
trafficability, peat harvesting, animal and crop disease models can all benefit from the
meteorological NWP output. Therefore it is essential that the agri-modellers know how
to access this information. As NWP output originates almost exclusively from national

meteorological services, the first step to obtain data is to discuss your requirements with the agmet, commercial or IT unit within an appropriate NMS.

8.5.1   Format (technical aspects) and local adaptation

Output from NWP is, in most cases, grouped in fields of grid-point values of a particular weather variable for a particular time-step. Generally the available timesteps are at 6-hour intervals, but individual NMSs may provide data at finer time resolutions or accumulations (e.g. for rainfall) or extrema (e.g. for temperature) during a time period. As the grid-points are separated by distances of the order of 30 km (at the time of writing), they do not normally coincide with the location(s) of interest. Therefore the user (or the supplier) will need to interpolate, introducing an element of smoothing of the values. This can be avoided by simply choosing the nearest grid-point. Small-scale local topography will also introduce errors. For example, near the top of even quite a small hill, rainfall and wind can be expected to frequently exceed model values, whereas in a small valley, the opposite would be normal, and in addition, night-time temperatures in valleys might often be considerably below model values.

In the case of some meteorological elements, bias can be reduced by compensating for the known difference between actual and model orography using standard multipliers. Such factors are more suited to long term average rainfall and should be used with caution when modifying forecast data. The following equation gives an indication of the approximate size of the corrections for rainfall prediction:

$$R_{f \text{mod}} \approx R_{fNWP} * \left[1 + \frac{(0.315 * R_{ann} - 119) * (Z - Z_m)}{R_{ann} * 100} \right]$$

$$8.5$$

Where $R_{f\text{mod}}$ is the modified value of the rainfall, R_{fNWP} is the model forecast value, R_{ann} is the annual total at a nearby low level station, Z is the height (m) above sea level at the point in question and Z_m is the height above sea level of the model orography at the same point.

While temperature can similarly be converted (using $T_{fNWP}-(Z-Z_m)*0.0064$), results in this case are even more unreliable in specific cases where, for example, temperature can rise with height in the atmosphere. Note also that while the forecast mean daily temperature will normally be reasonably approximated by the average of the 6-hour output time steps for the day, maximum–minimum temperatures will normally be higher–lower than the highest–lowest of the four outputs. One formula (Stull, 1995) to calculate a Celsius wind chill using V as the wind speed in m s^{-1} and T in ^{0}C is:

$$T_{wc} = 33.0 - \left(\frac{V - 2}{2} \right)^{0.21} * (33 - T)$$

$$8.6$$

Wind chill as used in agriculture must, of course, sometimes take other meteorological factors into account. For example, rainfall is also relevant in the case of wind chill on young lambs (see Chapter 7).

The wind output at 10 m above the surface from NWP is the mean predicted wind speed. Users are sometimes concerned with the likely gustiness. As an approximation, for rolling country with few windbreaks a factor of 1.7 can be applied to the mean wind to obtain an estimate of maximum gusts. Note also that most current models are not good at predicting local winds such as sea breezes or valley winds. Table 8.2 shows approximate ratios of mean wind at levels below 10 m to that at 10 m. The precise ratio depends on factors such as atmospheric stability and surface roughness.

Table 8.2: Approximate ratios of mean wind speeds at levels below 10 m to that at 10 m

Height above ground	2	3	4	5	10
Wind / wind at 10m	0.78	0.83	0.85	0.88	1.00

8.5.2   Availability

A widely used and compact file format for NWP output is known as GRIB (a guide to the GRIB code is available from the World Meteorological Organisation at ftp://www.wmo.ch/Documents/www/guide-bin-codes.wp5). Output can also normally be provided in ASCII text files containing individual values row by row over the model area for the particular parameter and time-step. In most cases, however, agriculturists will require NWP information for a single point or a very limited area and the simplest solution is, having stated requirements, to allow the NWP data provider to produce the required parameters for the required location.

ECOMET is an Economic Interest Grouping of NMSs from most European countries. One of its aims is to expand the availability of meteorological information within the ECOMET territory. Copies of the price lists of NWP products from ECOMET countries and the catalogue of ECMWF products are available from member-state NMSs. Copies of the larger User Guide to ECMWF Products and the ECOMET Catalogue are usually available in the respective member NMSs.

8.5.3   Format of NWP output

Up until recently, all available NWP output was from deterministic models, which produce a single value for a given parameter at a given location and time. It is well known that the major source of inaccuracy in forecasts is uncertainty in the initial state. Therefore, on occasion modellers will run their models numerous times from slightly different initial states. Methods exist to determine the sensitivity of the forecast model to the analysis in any particular area. Small analysis perturbations can be introduced at the more sensitive areas before each run of the model. The resulting ensemble of outputs gives a range of results for a given location and time (Buizza *et al*, 1998). The spread of these outputs will in some cases give an indication of the reliability of the deterministic forecast. It is also found that the average of these ensemble outputs is frequently more reliable than the single (non-perturbed) deterministic output in the four-to-eight day range. The various results also will enable probabilities of different results or ranges of values to be quoted. Sometimes results will fall into a small number of alternative scenarios (clusters). For example, 50 model runs might result in 40 cases (80%) of dry conditions with light winds, sunny spells and high pressure, whereas the remaining 10 cases (20%) might show westerly winds and rainfall. Probability of precipitation forecasts, derived by other statistical methods, are commonly quoted in the USA. Considerable work has been done in recent years on this ensemble forecasting method and it can be expected to soon become a standard tool of the forecasting agencies in their efforts to communicate uncertainties and probabilities in the forecasts and to extend the useful lead time of their forecasts.

8.6   Combining NWP and Climate or recent information

In the case where a customer requires meteorological advice beyond the range where the NWP output skill is known to drop off significantly, increased value can sometimes be provided by modifying the output to bring it closer to climatology. Also customers (e.g. in the agricultural area) will often have historical figures (sales or yields, for example) for previous years. By providing the same meteorological variables for the historical period in question in the same format as the forecast values, the forecasting agency can provide greater value to the customer. For example, if conditions in the current year-to-date have been very similar to the previous year and if the NWP output is indicating better growth weather in the final week before harvesting than existed last year, such information can have considerable value to the farmer. An analysis of the historical records of the customer and their correlation with weather parameters can produce a weather sensitivity analysis of the customer's operations which will permit forecasts of direct relevance (e.g. t ha^{-1}) to those operations, as opposed to forecasts of meteorological variables (such as temperatures, rainfall and sunshine).

Also, the usefulness of agricultural models which require meteorological information over a longer period than can be reliably forecast (e.g. 20 day data for grass growth models) and which today use only current and recent weather information can be improved by incorporating NWP. For example, in the case of the 20-day requirement, instead of running the agricultural model each day with the previous 20 days' data it could be run with 15 past days and five forecast days.

8.7   The future

As computer processor speeds increase, parallel processing becomes more commonplace and storage drops in price, the distinction between model types mentioned in Section 8.1.7 above will continue to become more blurred and ensemble runs will become more commonplace. More and better observations, further improvements in data assimilation and further improvements in the model themselves will bring further improvements in the forecasts. It is likely also that, in the not too distant future, output data will be fed directly to farm or regional decision support systems. The following are some specific developments which can be expected.

8.7.1   More frequent model runs

In practice, starting a new NWP run every six hours is satisfactory for most purposes. However, there is a customer demand for updates at more frequent intervals or at intermediate times. The increased automation of the forecast production process requires a more or less continually updated NWP output. Formerly, updating the NWP output by using the data that becomes available during the six-hour intervals between NWP runs was a significant part of the human forecaster task. It is likely that NWP models will soon be run routinely at hourly or three-hourly intervals. While these models may not be as sophisticated as the main model which will be run at six or twelve hour intervals, they can be expected to add significantly to the output quality, especially for shorter-range forecasts. Such developments will benefit the agricultural community in their short term decisions such as to spray or not.

8.7.2   Better temporal and horizontal resolution

As computer storage becomes cheaper and communications become faster the availability of hourly (or even single time-step) output of all forecast parameters will become commonplace. This will enable automatic forecasts with more precise timing of events (e.g. transition through critical temperature thresholds and onset or clearance of rain) and better specification of extremes (e.g. wind speed, temperature and hourly rainfall). While new assimilation methods mean more observations taken at times other than the formal start time of the model will be incorporated. Also, the introduction in 1998 of a new Rapid Update Cycle (RUC – 2) model in the USA is an indication of things to come. This model, on a 40km grid covering the continental United States, runs every hour, produces high-frequency updates of forecast information and is used primarily, at the time of writing, for forecasts out to 12 hours ahead.

Improved horizontal resolution (i.e. smaller spacing between grid-points) will mean better accuracy in forecasts for specific points, particularly for rainfall and wind. To some degree, such improvements will follow from the better specification of the orography. It seems likely that within a few years NWP with resolutions below 10km will be commonplace, with some at a fraction of this over a more limited area. Plans in the USA, for example, are to have a national 5km mesoscale model operational by the end of 2001.

8.7.3   Better use of observed data in NWP

Traditionally, NWP used only a subset of the observed parameters at the surface. Improved model formulations are expected to allow more observed parameters to be used in the initial analysis. These will include information on visibility and cloud height and type. Increased density of observations resulting from improved automatic weather stations and other technological developments will improve analyses (and consequently NWP output), especially in the case of mesoscale models.

Weather radar information is also likely to be incorporated in the initial analysis both for moisture/rainfall analysis and for wind at upper levels in the atmosphere (using Doppler radar).

While information derived from satellite images is currently used in NWP analyses, new satellite sensors permit pressure and surface wind observations over the sea to be remotely measured and provide improved upper air observations. The incorporation of such observations into the NWP initial analyses will improve accuracy in the extended range generally but is expected to bring significant improvements in the shorter range NWP over Ireland where much of our weather originates in the currently data-sparse Atlantic. A new generation of satellite-borne infrared atmospheric sounding instruments is expected to provide significantly improved temperature observations over a wide geographic area through the depth of the atmosphere. A good example is the EUMETSAT-sponsored Infrared Atmospheric Sounding Interferometer, scheduled for operations in 2003 (Collard, 1998). Finally, a new analysis method (known as 4-Dimensional Variational analysis) which is very demanding on computational resources will become commonplace. This should lead to improved results by enabling full use to be made of observations taken outside the nominal analysis time of the NWP run.

8.7.4   Combined climate data, current meteorological data and forecast data

Traditionally these sources of information have been treated separately. Including climate and current information with the NWP output will provide a more useful package to the customer. As PCs with sophisticated graphics packages become more commonplace, customers in the agricultural area can have the meteorological data presented to them in a more readily useable way or can have the data automatically input to their agricultural models to produce output that is directly related to their areas of interest. Thus, rather than being provided with temperature, rainfall, wind and humidity forecasts, the agricultural community can be provided with derived information such as rate of drying or with contextualised information such as forecast compared to normal or to last year or to normal frequencies ("next week is likely to be cold - as cold as would be expected at the time of year only one year in ten"). Direct agricultural advice could also become available - 'next week will be suitable for the spread of potato blight; tomorrow will be a suitable day for spraying'- or advice in relation to yield and quality of crop at harvesting - 'dry today and the following two days also expected to be dry - good for harvesting. In the absence of rain, the projected moisture content of grain crops should decrease from 40 per cent to about 30 per cent within five days, but there is an 80% probability of continuous heavy rain at that stage, increasing the moisture content again and making harvesting problematic'.

8.7.5   Neural Networks and NWP

While neural networks have been used in agricultural modelling (Elizondo et al., 1994), they also have use in the field of meteorology. Such networks are essentially computer models which 'discover' relationships between the inputs and the outputs as opposed to having such relationships prescribed within their coding. Use of direct NWP output as part of the input to neural networks or other artificial intelligence systems can be expected to bring further improvements to weather forecasts in the coming years. This is likely to find use particularly in the area of extreme or rare event forecasting and in modifying the NWP to take account of very local effects. For example, Kuligowski and Barros (1998) found the neural network approach provided significant improvements (over direct NWP output and other statistical processing methods) for localised prediction of moderate to high precipitation amounts. For a general overview on the subject of neural networks a text such as Russel and Norvig (1995) is informative. Koizumi (1999) describes a neural network method of producing improved precipitation forecasts using observations and NWP as input. Hsieh and Tang (1998) describe the application of neural networks in meteorology, although they do not deal with the specific case of using NWP data as input to a network.

8.7.6   Concluding remarks

Many of the above developments are already being employed experimentally and can be expected to lead to significant improvements in the quality of forecasts of surface variables in both the short and extended time range. Some NWP output variables which are generally considered experimental at the time of writing will become mainstream products. Amongst these are some items of particular interest to the agricultural community such as evaporation, soil moisture and radiation (in both the global and the photosynthetically active parts of the spectrum). In the next decade developments mentioned above and others can be expected to bring a one-day improvement in the useful forecast range (ECMWF 10 year Plan) while at the same time improving accuracy in the shorter range forecasts.

Just as atmospheric models and oceanic forecasting models are now beginning to be more closely coupled, in the longer term it seems likely that models will become much more than atmospheric models. They will include, for example, the complete hydrological cycle (with observations and forecasts of such things as evapotranspiration, ground water and stream flow) and detailed atmospheric chemistry (incorporating observation and prediction of atmospheric pollutants). As well as providing direct output of new elements of direct relevance to the farming community and others, the increased comprehensiveness of such integrated models will feed back to the quality of the traditional weather elements.

While this chapter has concentrated on direct model output of meteorological elements, the reader is reminded that statistical post-processing and modification by the skilled human forecaster still adds considerable skill to the forecasts. Having said that, there is no doubt that NWP is responsible for the huge strides in forecast skill during the second half of the 20th century and that NWP will continue to be the mainstay of weather forecasts well into the future.

References

Buizza, R., T. Petroliagis, T. Palmer, J. Barkmeijer, M. Hamrud, A. Hollingsworth, A. Simmons and N. Wedi. (1998): Impact of model resolution and ensemble size on the performance of an Ensemble Prediction System. *Quarterly Journal of the Royal Meteorological Society* **125(550)**: 1935-1960

Carson, D. J. (1999): Climate Modelling: Achievement and Prospects. *Quarterly Journal of the Royal Meteorological Society* **125(553)**: 1-28.

Collard, A. D. (1998): *Notes on IASI Performance*. UK Meteorological Office FR Division Technical Report No. 256.

ECMWF. (1997a): 1987 – 1997: Ten years of research and operational activities with the Integrated Forecasting System (IFS). *ECMWF Newsletter* **75**: 2-7.

ECMWF. (1997b): Seasonal forecasting at ECMWF. *ECMWF Newsletter* **77**: 2-8.

ECMWF. (1997c): *Verification of ECMWF products in Member States and Co-operating States*. Report, August 1997.

Elizondo, D, G. Hoogenboom, R. W. McLendon, (1994): Development of a Neural Network Models to Predict Daily Solar Radiation. *Agriculture and Forest Meteorology* **71**: 115-132.

Haltiner, G. J. and R. T. Williams. (1980): *Numerical Prediction and Dynamic Meteorology*. John Wiley & Sons.

Hsieh, W. W, B. Tang. (1998): Applying Neural Network Models to Prediction and Data Analysis in Meteorology and Oceanography. *Bulletin of the American Meteorological Society* **79**: 1855-1870.

Katz, R. W. (1987): On the Convexity of Quality / Value Relations for Imperfect Information on Weather or Climate. Preprints, *Tenth Conference on Probability and Statistics in Atmospheric Sciences*, 91-94.

Koizumi, K. (1999): An Objective Method to Modify Numerical Model Forecasts with Newly Given Weather Data Using an Artificial Neural Network. *Weather and Forecasting* **14**: 109-118.

Kuligowski, R. J., A. P. Barros. (1998): Localized Precipitation Forecasts from a Numerical Weather Prediction Model using Artificial Neural Networks. *Weather and Forecasting* **13**: 1194-1204.

Lorenz, E. N. (1963): Deterministic Nonperiodic Flow. *Journal of Atmospheric Science* **20**: 130-141.

Lorenz, E. N. (1982): Atmospheric Predictability Experiments with a Large Numerical Model. *Tellus* **34**: 505-513.

Puri, K., G. S. Dietachmayer, G. A. Mills, N. E. Davidson, R. A. Bowen, and L. W. Logan. (1998): The new BMRC Limited Area Prediction System, LAPS. *Australian Meteorological Magazine* **47**: 203-223.

Russell, R. and P. Norvig. (1995): *Artificial Intelligence–A Modern Approach*. Prentice Hall International Editions.

Simmons, A. J. (1996): The Skill of 500 hPa Height Forecasts. *Proceedings of the ECMWF Seminar on Predictability (Reading September 1995)* **1**: 19-68.

Snook, J. S., P. A. Stamus, J. Edwards, Z. Christidis, and J. A. McGinley. (1998): Local-Domain Mesoscale Analysis and Forecast Model Support for the 1996 Centennial Olympic Games. *Meteorological Applications* **13**: 138-150.

Stull, R. B. (1995): *Meteorology Today for Scientists and Engineers – A Technical Companion Book*. West Publishing Company.

Wilks, D. S. (1997): Forecast Value: prescriptive decision studies. In R. W. Katz and A. Murphy (Editors): *Economic Value of Weather and Climate Forecasts*. p 109-146. Cambridge University Press.

Chapter 9 An endnote: The future for models and modelling in agro-meteorology

N. M. Holden and The Agmet Group
Department of Agricultural and Food Engineering
University College Dublin, Ireland

9.1 The role of the model in translating data into information

9.1.1 Information, its value and dissemination

Many challenges face the agro-meteorological community over the next decade. Foremost among these challenges are changes in the process of disseminating information, and meeting end-user requirements for new kinds of information. Problems lie in locating the appropriate product (e.g. raw or processed data, models and management tools), in using the product (e.g. issues of computer and software compatibility) and with the "gap of access" (a term suggested by Eric Benhamou, CEO and Chairman, 3Com Corporation, speaking on 6th January, 2000 – that is, the difference between those who can afford access to technology and those who cannot). Data are the basis of agro-meteorological information. Data of use to the agro-meteorologist arise from both the environment (characteristics of the earth's surface), and the weather (characteristics of the earth's atmosphere). Having quantitative data for variables such as incident radiation or rainfall is of more value than merely knowing that it rained, or was sunny, if there is some means of interpreting the data, i.e. a model. Quantified information also aids prediction; this is the route to estimating what the future holds.

Raw data are of relatively little use to the farmer, or other land managers. Data flows via researchers, meteorologists and agro-meteorologists, who filter and interpret the raw numbers and making some sense of them (Figure 9.1). Ideas and concepts are combined in models to produce weather forecasts and agro-meteorological forecasts. Thus data becomes useful information of great value. Television, radio and the press (i.e. the media) are the main vehicles that convey information to the end user.

Agricultural advisers provide a vehicle for translating research into practice on the land and for assisting in the interpretation of the information available to the land user. Quantified models have a role to play in the utilisation of information. Once models are developed to the point that they can be of use as decision support tools, the end user can have automated support for management activities by means of existing communications technologies. Increasingly the Internet and world-wide-web sites are being used to disseminate information to the end user, but these routes are dependent on end-users adopting the technology. The case of mobile telephone ownership in Ireland illustrates this point. About 80% of farmers use mobile telephones, but perhaps

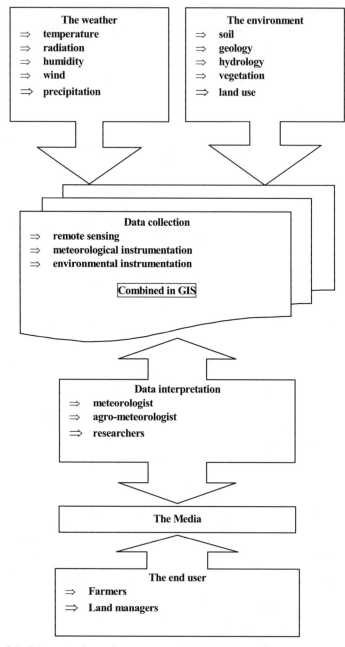

Figure 9.1: Schematic flow of agro-meteorological information

only 15-20% use an Internet connected PC. This means that a service delivered by text message is probably going to be more useful than via the Internet provided there is adequate coverage in all regions. TV ownership is even greater than that of the mobile telephone, so teletext services, though seeming dated, should be a primary source of information for the farmer.

One advantage of the internet and multiple data sources is that in the future, a land manager can choose the route between the raw data and the user based on how the data are analysed. The value of data is only as great as the quality and the usefulness of the model used to process it and turn it into useful information. The manager will have to choose the system that provides the best information for the task at hand. The role of agricultural advisors in the information chain will also increase as they gain more access to well-designed and calibrated models. The model provides the advisor with ideas and concepts that may not be readily available locally, and a means of testing ideas without endangering anyone's livelihood. In the future the farmer may have to decide where to invest money for advice: in a person (the agricultural advisor) or in a decision support system (a computer model).

9.1.2 Models as translators and consolidators

With the increasing complexity in modern farming and developments in meteorology, it is necessary to condense data into simplified information. This book highlights some of the vast amount of available data (Figure 9.2). In order for these data to be of any worth they have too be made useable and manageable. The model is an ideal means of adding value to data and filtering or consolidating to the right degree for the end user (Figure 9.3). For example, in order to initialise and run numerical weather prediction models there are vast amounts of data required. However, the end user might only want to know about a few basic variables for one location. Thus the numerical weather prediction model has translated and consolidated the original data into useful information – predictions of future weather – for the end user.

Effectively it is the model that makes the raw data valuable. The value of the output data from the model depends on two key features: the quality of the input data and the quality of the model. In Chapter 5 we evaluated three approaches to the same problem, i.e. predicting grass yield, with a view to assessing both the principles of the models presented, and the characteristics of the output. Each of the models translated the data presented into a more valuable output (predicted grass yield). The question that the end-user has to ask is: are the results reliable and accurate? If not, the model has added no value to the raw data used in the prediction process. Comparative evaluation of models by research and advisory staff is to be encouraged in order to ensure that end-users get the best value from the data available.

Data from national meteorological agencies are subject to quality control and cross-checking and can be regarded as being of the highest available quality. The same cannot be said of data from someone's back garden. The adage rubbish in - rubbish out holds true. Poor quality, worthless data cannot be made valuable by a model. Equally, if good quality data is fed into a poorly conceived model, then the result will also be of little worth. The principles of model development and experimental design are very important because they dictate the worth of the information that comes out of a model. If the model is conceptually flawed (e.g. an assumption is not tenable), or inappropriate mathematics have been used then the result is meaningless. There may be more compact and easily assimilated information, but its value will not have increased.

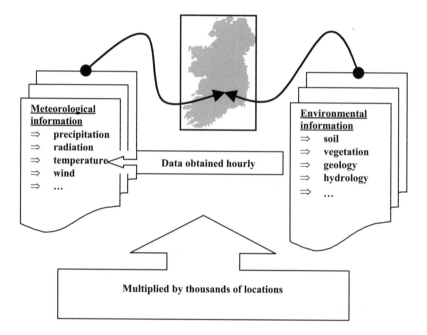

Figure 9.2: The quantity of information generated to describe just one site is too much to use unless it is filtered, translated and compacted

9.1.3 Value added information

Regardless of how valuable information is, and its currency in the modern world, more and more information is becoming available as time goes by. It is now possible to receive images from meteorological satellites on a personal computer within a short time of their being captured. There are many sources of weather data via the Internet, telephone services, television text systems and the media. With the development of more specific models aimed at particular end users, the information that is available will become more user-oriented and user-specific. It is this process of tuning information to the user that adds to the value of the data available. User specific tuning is dependent on models, therefore it is going to be increasingly important for agro-meteorologists and other environmental scientists to develop and use models that will eventually become part of the management of data flow from the source to the target.

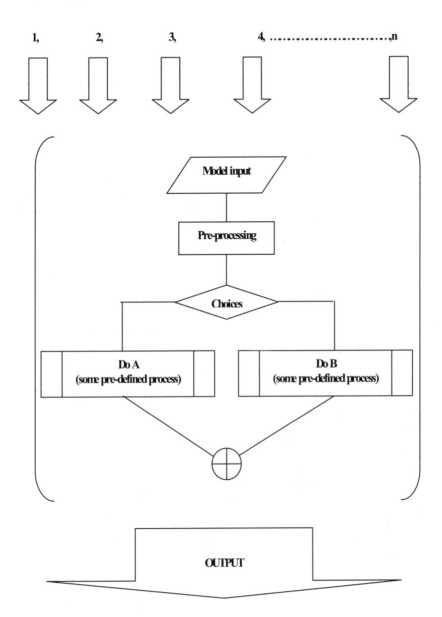

Figure 9.3: Schematic representation of a model as a data translator, filter and compactor resulting in added value of information

9.2 Future challenges

9.2.1 Scale, applicability, reusability and testing

Most models are built to describe a particular scale of operation, be that micro-, meso- or macro-scale. The reasons for this are: the scale at which the processes being modelled operate, and the scale at which the mathematics in the model operate (based on underlying assumptions). It is desirable that in the future, modelling tools allow the user to work at any scale required in an integrated manner. Additionally, the user should not be responsible for selecting the correct scale when applying a model. In practise a tool to assist a farmer in management tasks should provide information about the **forthcoming weather** (i.e. macroscale, ranging from national through regional to catchment scale), specifics about the **water balance** and **environmental conditions** on the farm (i.e. catchment through farm and field scale right down to plant scale), and **microclimate effects** of importance (i.e. field to plant to sub-plant scale effects such as differences in soil temperature influencing germination or small scale shading effects influencing establishment of young plants). Figure 9.4 schematically represents the scales of interest in agro-meteorology that are of use in agricultural production and environmental protection schemes. Currently this type of tool only exists in the farmer's brain where personal experience handed down through generations enables the farmer to "know" what the consequences of a particular weather forecast might be. Such knowledge is effectively being reinvented by researchers in precision agriculture who are trying to recreate management on a sub-field scale that is analogous to the smaller fields that were combined to form large modern fields.

Effective agro-meteorological models must be incorporated into decision support systems that describe the whole farm environment at a range of scales from 1 m to 1000 km (if we consider the scale of transport of livestock and products between individual land holdings). If a decision support system (DSS) is to be useful, it will either need a model that crosses scale boundaries, or a series of models that describe the same phenomenon at different scales. The example in Chapter 6 used different models of runoff depending on the availability of data. In the latter case, the DSS will have to select the correct scale of model to use. This is likely to be a more common scenario because there are few conceptual ideas that can be applied with assumptions that remain tenable over a wide range of scales. The notion of a DSS with many models all working to help the farmer is a utopian dream simply because of the vast amounts of data that would be required to calibrate the models. Additionally, the models would need data for testing and subsequent application; all this adds up to vast amounts of data for just a single location.

At some stage, the notion of scale becomes entangled in the concept of geographical extent over which a model is applicable for a single set of calibration data. If a regional model is calibrated for a country, does a farm model need calibrating for a region? Perhaps more importantly, will each field model need calibrating, and each plant scale model? If we consider the latter case, we end up needing so much data that the whole idea becomes a nonsense. Thus the scale at which a DSS could operate will be limited by the practical reality of obtaining data to use with it. If we consider the value of data, the cost of running a DSS on a farm could become prohibitive unless the pricing of data enables a balanced cost-benefit outcome.

Scale Hierarchy of Models

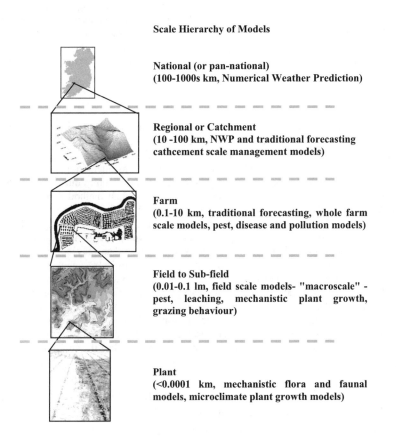

National (or pan-national)
(100-1000s km, Numerical Weather Prediction)

Regional or Catchment
(10 -100 km, NWP and traditional forecasting cathcement scale management models)

Farm
(0.1-10 km, traditional forecasting, whole farm scale models, pest, disease and pollution models)

Field to Sub-field
(0.01-0.1 lm, field scale models- "macroscale" - pest, leaching, mechanistic plant growth, grazing behaviour)

Plant
(<0.0001 km, mechanistic flora and faunal models, microclimate plant growth models)

Figure 9.4: Scales of modelling that need to be integrated in the future

These ideas raise the issue of data being suitable for multiple uses and being reusable. The same soil data could be used to initialise, calibrate and test models for leaching, gaseous emissions, runoff and plant growth, but only if the models are developed in such a way as to operate at similar scales and with similar theoretical descriptions of the soil. As the role of modelling in agricultural science increases and becomes part of everyday agriculture, it will be necessary for models to integrate in terms of both data requirements and data outputs. GIS tools will go a long way in assisting this process.

In view of the fact that models are required to provide land managers with a set of management tools, the question of model testing arises. The need for calibration is paramount, but just as importantly models need to be tested (the term "verified" is also used). Following the principles outlined in Chapter 1, data are required to build a model, independent data to calibrate it, and yet more independent data to test it! All this is required before practical application. If this crucial, but time consuming, testing stage is omitted then modelling will receive a poor profile in the media as unsuitable decisions are taken on the basis of a poorly developed model. Once a reasonable model of part of the agricultural system is developed, it should become a priority for other researchers to test the model, rather than build a competing model as tends to be the case at the moment.

9.2.2. Internationalisation

Transfer of models from their point of origin may be limited not only by the scale of application, but also by differences internationally, in terms of environmental conditions (effecting boundary conditions in a model) and even methods of farming (e.g. degree of mechanisation). Keane et al., (1998) suggest that obstacles to international transfer of models include: (i) errors in primary input data (observations), (ii) bias introduced in the processing of the data, (iii) use of inappropriate scale (original method derived for a particular scale), (iv) uncertainties in the ecophysiological parameters calibrated in the model, (v) interpolation techniques used locally and (vi) unaccounted effects of pests, diseases and management decisions. Applying a model developed with a robust network of meteorological observation sites in a country with sparse data may be impractical, and could lead to seriously erroneous results. Regression models cannot be applied internationally using locally derived co-efficient values without rigorous re-calibration. This of course, is not always possible, but is very important because of differences in environment, and the representitivity of the observation locations available at the source of the models, and the location of application. Many of the regression type models presented in Chapter 7 for instance, are useful in raising awareness for end-users internationally, but the actual numbers used in the equations reflect the environment and observations stations used in their development.

9.2.3 Data standards and unified approaches to data acquisition, management and dissemination

In this book we have largely focused on Irish data due in the main to the expertise and national perspective of the authors. With the increasing need for data from larger area (such as for NWP, Chapter 8) there is a need for standard data types. A user relying on data (particularly if purchased) should be able to expect an agreed standard. If a model is to be applied in another country, the same standard of data needs to be available. The example of not mixing sunlight hours and radiation values in a model (Chapter 3) illustrates a simple problem of this type. In Chapter 2 the issue of "support", in the geostatistical sense (i.e. the area or nature of a site represented by a measurement) is a somewhat similar concept. Data representing the same variable should be comparable in order that models be transferable between locations.

Jones and Thomasson (1998) drew a number of conclusions with regard to harmonisation of data in the EU for land management, many of which are generally applicable to environmental management at large. If a model tool is developed in a country with a dense data network, it may prove difficult to use in a country with sparse data, or the quality of the output may be compromised. Thus in the future there is a need for agreed standards of data in terms of (i) measurement method, (ii) quality control and (iii) spatial and temporal sampling resolution. In order for data to be widely disseminated and useable it should always be in a form that is accessible from all current computing platforms, and flexible enough to be accessible in the future. An example of a practical problem in this regard is archives of satellite imagery (useful for change detection research) are available on large spool computer tape only. However machines to read such data are no longer widely available in modern computing environments. Data storage technology does not always advance with changes in computing standards! It is of particular importance that those who hold data make the availability known to potential users. As can be seen from Chapter 3 and 4, there are data available but they have to be searched out and bought together by those developing models. Integration from many data sources for a routine management task is much more difficult. It is necessary that the procedure for acquiring data in the future be clarified and simplified, along with the format and resolution.

9.2.3 Encouraging the adoption of models for land-use management

If agrometeorological models are to become an integral component of on-farm decision making, they must gain general acceptance in day-to-day farming and land-use management practise. Success along these lines is most likely to occur if agrometeorological forecasts are part of a farm expert system. Problems that currently arise in this area are exemplified by yield potential prediction. In general the yield potential indicated by models exceeds the actual yield value. This is caused by limitations in the prediction of the accumulation of dry matter during specific phases of growth which are not correctly accounted for by the model. Meteorological phenomena responsible for unexplained reduction in yields include sudden spells of unseasonably low temperature or frost, high temperature, excessive or insufficient rainfall and excessive windspeed. The critical short-term conditions for significant yield reductions may depend on crop stage, whether it is at a pre-establishment, establishment, vegetative, flowering, ripening, harvest or post-harvest phase. For the model to be a useful component of an expert system data with a temporal scale (and possibly spatial scale when considering precision agriculture) will be required to account for non-macroscale phenomena. The appropriate threshold values for the meteorological factors, and their combinations, which can cause significant yield reductions, need to be determined locally. Typical situations which cause stress in crops include longer-term periods of excessive evaporative demand, extended canopy wetness, inundation, severe lack of moisture or excessive heat. To be useful in assisting land management, additional information is required for agro-meteorological models. Details relating to field preparation, sowing or harvesting, under either adverse or appropriate agricultural conditions (e.g. delayed due to excess wetness; harvest not hindered by rain), crop protection measures (higher than average attack of pests and diseases caused by certain weather conditions at critical times), or the occurrence of temperature and water extremes (abundant spring rains could have a negative effect in terms of nutrient limitations, i.e. maize, vegetative to shooting stage) need to be known. An interdisciplinary approach, as reflected in the expertise of the contributors to this book will be necessary to get the best from the models of the future.

The establishment of a local crop knowledge-base, specifying situations in which crop yield could be significantly reduced below the potential set by climate needs to be undertaken (Russell and Muetzelfeldt, 1998). Such a crop knowledge matrix could be constructed from observed, and sometimes anecdotal data accumulated by crop experts over several years. Setting up a computer-resident, crop-knowledge base is an appropriate undertaking for research institutes or advisory services, possiblly in consultation with universities to enhance day-to-day routine use of agro-meteorological models.

9.3 What has the content of the book told us?

Having considered the principles, data, applications, and briefly, the future of modelling of agro-meteorological applications, what have we learned? It is possible to summarise the lessons learned succinctly as: (1) *Model output is only as good as the effort that went into making the model*. It is essential to follow the basic principles of modelling to ensure a good model is produced (Chapter 1); (2) *Output from a model is only as good as the data that goes in*. Strive to have good data. Do not cut corners by estimating values or using convenient numbers unless absolutely necessary, and then use the best possible methods (Chapters 2, 3, and 4); (3) *Understand what the model is doing*, the assumptions associated with the theory and mathematics, and the wider system about which the model is a simplified component (Chapter 5); (4) *Models can be tools to help us understand the world, but also tools to help us apply responsible management*. A model has to be used correctly in order to be of any use what-so-ever (chapter 6); (5) *A multi-disciplinary approach to future developments is the best way of making progress*. This book should help many readers to understand how the various specialisations fit together to make a whole; and (6) *The construction of a model can help reveal weaknesses in available data, and in the knowledge or understanding of the model developer*. Of course, all these conclusions apply to any type of scientific system modelling, but due to the importance of agricultural land management with respect to the quality of our environment, there are of particular importance with respect to the subject of this book.

References

Jones, R. J. A. and A. J. Thomasson (1998): Summary of Progress and Recommendations. In: Heinke, H. J., W. Eckelmann, A. J. Thomasson, R. J. A. Jones, L. Montanarella and B. Buckley (Editors) *Land Information Systems. Developments for Planning the Sustainable Use of Land Resources*. European Soil Bureau, Research Report No. 4. p72-74.

Keane, T., G. Russell, M.N. Hough and D. Rijks (1998): Alarms. In: D. Rijks, J.M. Terres and P.Vossen (Editors) *Agrometeorological Applications for Regional Crop monitoring and Production Assessment*. Joint Research Centre, European Commission, EUR17735 EN, Ispra, Italy. pp 199-206.

Russell, G. and R.I. Muetzelfeldt (1998): Crop knowledge base. In: D. Rijks, J.M. Terres and P.Vossen (Editors) *Agrometeorological Applications for Regional Crop monitoring and Production Assessment*. Joint Research Centre, European Commission, EUR17735 EN, Ispra, Italy. pp 207-224.